AutoCAD for Windows Express

Tim McCarthy

AutoCAD for Windows Express

With 195 Figures

Springer-Verlag London Ltd.

Timothy J. McCarthy, PhD
Lecturer, Department of Civil and Structural Engineering,
UMIST, P.O. Box 88, Manchester M60 1QD, UK

ISBN 978-3-540-19865-9   ISBN 978-1-4471-2084-1 (eBook)
DOI 10.1007/978-1-4471-2084-1

British Library Cataloguing in Publication Data
A catalogue record for this book is available from the British Library

© Springer-Verlag London 1994
Originally published by Springer-Verlag London Limited in 1994

Typesetting: Camera ready by author, using LaTeX.

69/3830-543210 Printed on acid-free paper

To Grace and Fiona

ACKNOWLEDGEMENTS

I would like to express my deep gratitude to all those who helped me with this project. I am indebted to my colleagues in the Department of Civil and Structural Engineering at UMIST. My thanks go to Adel Nasser for his patient assistance with LaTeX and to David Walls and David Riley for support with the hardware. Thanks also to my students from whom I have learned so much and to the Autodesk team in Guildford.

My special thanks to Grace and Fiona for everything.

April 1994   Tim McCarthy

Trademark acknowledgements

ADI, AME, AutoCAD, Autodesk and DXF are registered trademarks of Autodesk, Inc. Postscript is a trademark of Adobe Systems. MS-DOS, Windows and Microsoft are trademarks of the Microsoft Corporation. Hewlett-Packard is a trademark of Hewlett-Packard, Inc.

CONTENTS

Chapter 1 INTRODUCTION

What is AutoCAD?

AutoCAD is the world's most popular computer-aided drafting package for the personal computer (PC). It is a fully functional 2D and 3D CAD program. Full 3D wire frame representation was incorporated in the program with the launch of Release 10 in 1988. Release 11 brought additional 3D facilities including some solid modelling capabilities. These capabilities were enhanced with Release 12 of the program for DOS and Windows. Its popularity has made AutoCAD the de facto industry standard for PC-CAD with a host of other program developers providing application software conforming to the AutoCAD format.

   As a fully functional drafting program, AutoCAD can achieve anything that can be drawn on a drawing board. The main benefits of CAD come more from being able to edit and exchange drawing information rapidly rather than simply replacing the drawing board. Starting to use AutoCAD is a difficult step as it requires a certain amount of new skill development. Once you have made the commitment to learn how to use the program and implement it in your everyday work the benefits will soon accrue. You will quickly discover that there are many things that you can do with AutoCAD that you could never do with a drawing board.

   With AutoCAD your drawings become more than just black lines on a white sheet of paper. The AutoCAD drawing is a database of information. Some of this is indeed graphic information, but AutoCAD knows the length of every line on the drawing. It knows what symbols and parts have been included on the drawing and it can output this information to design programs of spreadsheet programs for bill of materials and cost analysis.

The aims of the AutoCAD Express

The main aim of this book is to introduce AutoCAD users to effective CAD drawing techniques. This is done through structured exercises that demonstrate the AutoCAD drafting principles clearly. The commands are dealt with in this context as tools to make the job easier. It is also hoped that you will have fun doing these exercises and creating some of the pretty pictures.

The AutoCAD Express is suitable for new users as it covers the program from the very basics right through to advanced techniques. Occasional users will find it a useful and quick refresher, while even seasoned users will discover novel aspects to old commands. Not only are the commands fully described but AutoCAD *drawing techniques* are explored with many examples.

This edition of the book has been written for AutoCAD Release 12. While the main emphasis is on the Windows version the book fully supports Release 12 for DOS. It covers all the important aspects of the version with full descriptions of the 3D functions and dynamic viewing. The operating system used is Microsoft Windows and DOS. Where Windows specific functions are explained, a DOS equivalent is also given. References to the operating system are kept to a minimum, so users of OS2 and UNIX should not be distracted.

Because it is so flexible, AutoCAD can seem unwieldy to the new user. The exercises in this book follow each other logically along a well defined learning curve. Each chapter represents a stage along this curve, and at each stage the user can pause to consolidate the skills obtained, or proceed to the next stage. To overcome the sheer size of the AutoCAD program and the number of facilities available, the user is directed through the most appropriate path to complete the example drawing. You will never be overwhelmed by lengthy descriptions of abstract concepts and myriad command parameters. Rather you will learn things when you need to know them. By the end of the book there will be little left about AutoCAD that you will still need to know.

The Express route through AutoCAD

The AutoCAD Express is designed as a tutorial guide to the varied facets of the world's most popular computer aided drafting package. The emphasis is on *doing* the various commands and *achieving* results. Chapters 2 to 8 each present instructive drawing exercises which call on AutoCAD's drafting facilities in a logical order. Each chapter has a broad theme with useful asides included where appropriate. Each new command and facility is introduced in the context of solving a particular drafting problem.

Chapter 2 provides a quick introduction to the essentials of producing a drawing file. Chapter 3 gives a complete description line of drawing in Auto-CAD with detailed examples of the User–AutoCAD interface. In Chapter 4 the Express takes to the skies to introduce the bulk of AutoCAD's drawing commands. Your first encounter with the program's editing facilities also happens in this chapter. In Chapter 5 the AutoCAD Express lands in Paris to explore more advanced editing features and construct Gustav Eiffel's famous tower. It's back to the steamy kitchen for Chapter 6 where you will learn how to make and manipulate AutoCAD blocks and create a library of symbols. These symbols are used to help AutoCAD Express Kitchens Ltd. to quickly design

new fitted kitchens with automatic bills of materials. Their competitors must
be worried! Chapter 7 covers automatic dimensioning and a few other high-
level commands. The world tour continues in Chapter 8 from the unlikely start
back in the kitchen. This particular leg of the journey covers isometric projection and a 2.5D view of the Big Apple before visiting the pyramids of Giza in
glorious 3D colour. All the major new facilities introduced for 3D drafting are
described with relevant examples covering 3D drawing and visualisation.

Chapter 9 deals with aspects of printing and plotting your drawings and
how to get the right output at the first attempt. This includes a brief description of the Windows Print Manager and communicating between AutoCAD
and other Windows applications.

Finally, there are a few appendices covering technical aspects of Auto-
CAD. Appendix A describes how to configure AutoCAD on a new computer.
Appendix B provides useful hints on loading applications and some of the "ex-
tras" that come with AutoCAD. You will find a comprehensive index at the
back of the book.

Conventions used in the AutoCAD Express

The style of presentation is fairly simple. There are no distracting icons or
hieroglyphics. Plain English is used throughout and where jargon cannot be
avoided it is clearly explained. There are a few computerese phrases used in
the text which have helped me in writing the book and, I hope, will help you
in reading it. Here they are:

RETURN or ENTER?

These are two words that mean the same thing. AutoCAD will frequently tell
you that you must "Press RETURN to continue". Now, most keyboards don't
have a key called "RETURN" but do have one with "ENTER" or one with a
right-angled arrow. All three mean to "enter" the line by pressing the key and
"return" the cursor to the left margin. In this book the symbol <ENTER>
has been used to signify this. You will also find references to <SPACE> in the
book which mean "hit the space bar", gently!

Presentation of user–AutoCAD dialogue

What you have to type is shown in bold text. The AutoCAD prompts are
shown in normal text. Some points are referenced in diagrams and in the
dialogue. These references are presented in the text in brackets to the right of
dialogue. For example:

Command: **LINE <ENTER>**
From point: **35,40 <ENTER>**   (V)

This means that AutoCAD will display the word "Command:" and you have
to type the word "LINE" followed by pressing the ENTER key. AutoCAD
will reply with the prompt "From point:" to which you reply by typing the
two numbers and pressing the ENTER key. The "(V)" indicates that this
corresponds to the point marked with a "V" on the nearby diagram. You
should not type the "(V)".

Control keys

One of the keys on the PC keyboard has "CTRL" written on it. On some
keyboards the word is spelt out in full, "Control". When this key is held down
simultaneously with other keys special computer commands are executed. AutoCAD uses the control key in conjunction with a number of letters to execute
different commands. These are given in the text as, say, "CTRL B" or "^B".
This means to press the "Control" key and while holding it down also press
the "B" key.

Menus

From time to time Autodesk issues improved screen menus. With each new
issue, the display details change. Because of this, there may be small discrep-
ancies between the screen menus displayed in this book and those that appear
on your screen. The menus used in the AutoCAD Express are those distributed
as the Release 12 International English ACAD.MNU in the United Kingdom.

Other conventions

Some of AutoCAD's commands require more care than others. Those commands where errors can give disastrous results are preceded by *"HAZARD
WARNING!"*. Less dangerous commands are accompanied by a brief "Warning!". Don't avoid these commands. Just follow the safety procedures given
with the warnings.

Most pointing devices have more than one button. The "pick button" can
be found by trial, though it is usually the left hand one. Pressing the right
hand button in AutoCAD is the same as pressing <ENTER>.

The final note is not about jargon but is timely advice. Be careful not to
confuse 0 (zero) with the letter O (Oh), and 1 (one) with l (lower case L).

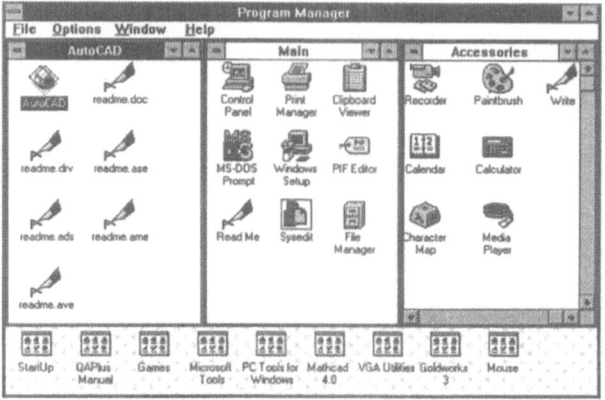

Figure 1.1 Program Manager Window

The Windows environment

Program Manager

If you are not already familiar with using Microsoft Windows then a brief introduction should help. Experienced windowers can leap straight to Chapter 2. What follows is a bare bones tour of Windows and is not meant as a substitute for reading the manual.

When you start Windows you are usually confronted by the Program Manager Window, similar to one shown in Figure 1.1. The details of what is visible on your system will depend on what applications you have and how many windows are open. It is from the Program Manager that all applications are loaded.

Along the bottom of the window shown in Figure 1.1 are most of the Program Group icons. To open up a program group window, move the mouse pointer (which should appear as a small arrow) on to the icon and double click the mouse button. The "double click" is two presses of the mouse button in rapid succession. Some program groups are already opened in the figure.

The "open" program groups shown in Figure 1.1 are: AutoCAD, Main and Accessories. Main contains two vital programs for use with AutoCAD for Windows. These are the Print Manager and the File Manager.

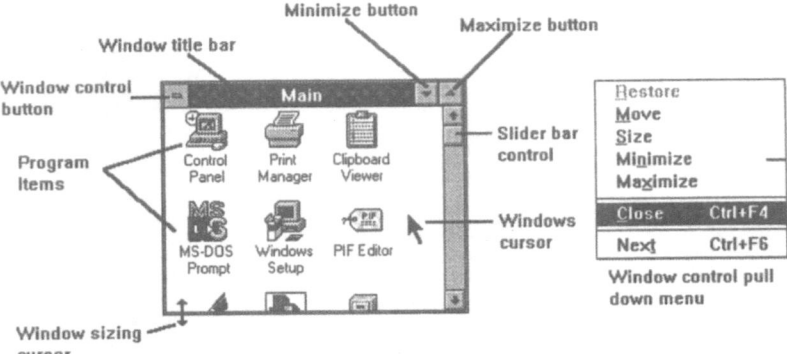

Figure 1.2 Window controls

Window control

The main features of any window are demonstrated in Figure 1.2. At the top of the window and in the middle is the title bar. To the left of this is the control button. Picking the control button gives the pull-down menu shown in the right-hand diagram. The **Close** allows you to exit the application and close the window. This menu also has options to minimize and maximize the window. Minimize means to reduce the window to a mere icon but without exiting the application. Maximize means to fill the whole screen with this window. To the right of the title bar are two buttons with triangle icons. These also minimize and maximize the window.

The program items are shown as small pictures or icons in the window. They are executed by moving the cursor onto them and double clicking. You can move an icon around the window by moving the cursor onto it and keeping the mouse button pressed drag it to a new location. Release the button when you are happy with the new location.

On the window shown in Figure 1.2 some icons are disappearing below the bottom of the window. They can be exposed by dragging the slider bar control down at the right-hand side of the window. Alternatively, the window can be enlarged. If you move the cursor slowly over the frame it changes into a double headed arrow. This is the window sizing arrow. Now press the button and keeping it pressed drag the frame down, then release. A window can be moved by picking the title bar and dragging it to a new location.

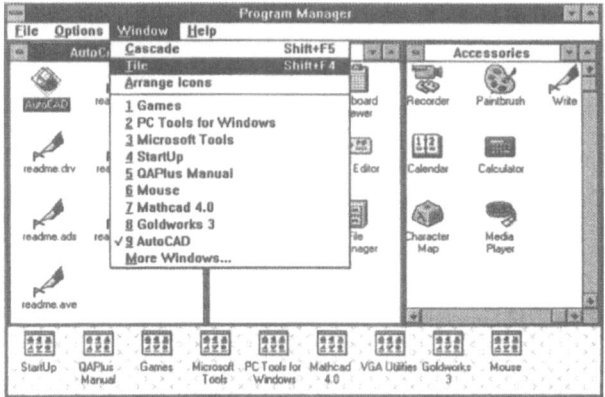

Figure 1.3 Window Pull-down menu

Window cleaning

Because the windows can be modified or moved with ease the screen can quickly become untidy. Indeed is it possible to lose windows off the screen or beneath other windows. This is where the **Window** pull-down menu comes in, Figure 1.3. This is activated in one of two ways. Point at the word, Window, and click the mouse button or use the keyboard. The underlined letters of the pull-down menu titles indicate the "hot key" combination to activate them. The combination is the ALT key and the underlined letter together. In this case press **ALT W**.

The **Tile** option tidies things up. If you look down the menu in Figure 1.3 you see all the program group names. There is a tick beside group 9, the AutoCAD group, indicating that this one is the currently active window. The title bar of the AutoCAD group window is shown in reverse also indicating its active status.

Windows hot-keys

You have already seen some of the hot key combinations for Windows operations. Here are a few more useful ones:

When more than one application has been loaded you can toggle from one to another by ALT and the TAB key. CTRL and the ESC key gives you a list of all currently active applications and allows you to switch to one of them or to end one. The F1 key invokes the HELP facility.

Enjoy the book and soon you will be enjoying the benefits of productive AutoCAD drafting!

Chapter 2 STARTING AUTOCAD

Preparation

This chapter assumes that either Windows 3.1 and AutoCAD for Windows or AutoCAD DOS Extender have been installed on the computer and are ready to be used. If this is not the case you can follow the procedure outlined in Appendix A of this book. Most dealers will install the software for you when they deliver it. It is good practice to request this service so that your purchase can be fully tested while the vendor is present.

Clear your work area so as to give comfortable access to the computer, keyboard and mouse. An area of about 200mm by 200mm (8in. by 8in.) should be adequate for the mouse. Switch the computer on and wait for it to go through its automatic self test and "booting" procedures. This will take approximately 30 seconds. You are now ready to start computer-aided drafting.

Starting AutoCAD for DOS

If you are using AutoCAD for DOS then you need to locate the startup file. This is usually called ACAD12.BAT or some similar name. If this is in the directory, C:\ACAD, then type **C:\ACAD\ACAD12**. This file sets a number of environment variables and then runs AutoCAD. Some people configure their PC to set the environment variables in the file, AUTOEXEC.BAT, which is run automatically when the PC is switched on. In this case you just need to locate the program file, ACAD.EXE. If this is in the C:\ACAD directory then type **C:\ACAD\ACAD**. Now go to the section "Creating a Drawing".

The AutoCAD for Windows icon

When the computer has booted, start up Windows. If it is located in a directory called "C:\WINDOWS" this can be done from DOS by keying **C:\WINDOWS\WIN** followed by **<ENTER>**.

In Windows you should be able to see the AutoCAD icon. If so then move the cursor arrow tip to the icon and click twice quickly with the left-hand button. After displaying a brief message the AutoCAD window should resemble that shown in Figure 2.1. Now proceed to the next section "Creating a

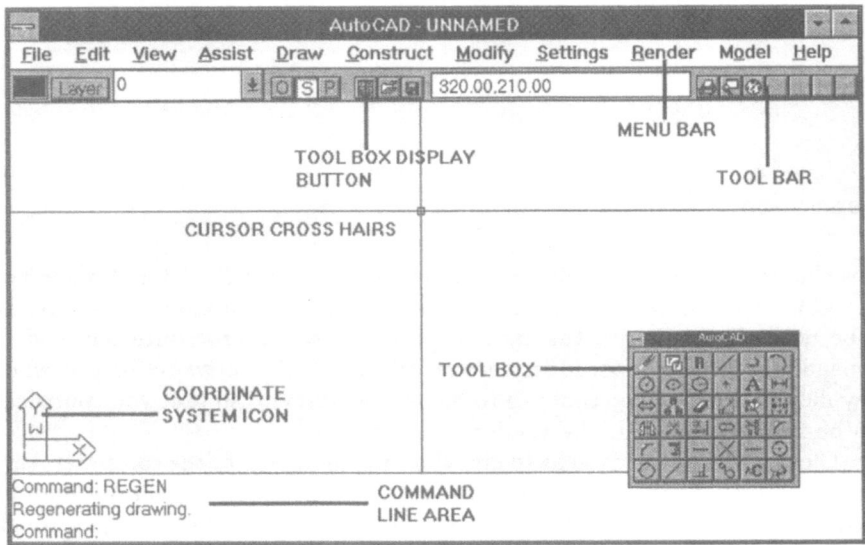

Figure 2.1 AutoCAD drawing window

drawing". If you get an AutoCAD window that looks drastically different from Figure 2.1 then check the "Menu preferences" section later in this chapter.

If you cannot see the AutoCAD icon when you start up Windows don't worry. The following sequence should help you to find it. Click on the Windows Program Manager icon (if it is not already active). Then click **Maximize** to make it fill the screen. You should now see the AutoCAD program group icon. Click on it and Maximize it. Finally, double click the AutoCAD "shell" or "drawing" icon and go to the next section.

If the AutoCAD icon still eludes you try picking "Arrange Icons" from the Windows View pull-down menu. This brings back icons that might be off the screen. Appendix A gives guidance on creating the AutoCAD program group and icons if you're still stuck.

Creating a drawing

The drawing window

Starting at the top of the AutoCAD window you first see the title bar which initially contains the words "AutoCAD - UNNAMED". This, "unnamed" will change to the drawing name as soon as one is defined. At the left of the title bar is the Windows Control menu icon. This allows you to close the window etc as described in Chapter 1. At the other end of the title bar are the minimize (triangle pointing down) and maximize (triangle pointing up) buttons.

Immediately below the title bar is the menu bar. There are eleven pull-down menus from File through to Help. These contain most of the AutoCAD commands and are described in detail throughout the book. Under the pull-down menus you will find the AutoCAD for Windows "tool bar". This line of icons contains a number of buttons which provide quick access to certain commands. The tool bar also contains a display of the coordinates of the cursor location. In the DOS version the tool bar is replaced by a status line giving the layer name and coordinates.

The bulk of the AutoCAD window is taken up with the active drawing area. This extends from the tool bar down to the command line area at the bottom. In the drawing area, the cursor is shown as a pair of cross hairs intersecting at the cursor location. The way the cursor is displayed changes with the context of the operation. When the cursor is moved into the tool bar or pull-down menu area it changes into the Windows arrow for clicking items.

In the drawing area you will also find the AutoCAD tool box. This is a floating button menu. Each of the icon buttons represents an AutoCAD command. As you move the cursor over the buttons the name of the appropriate command will appear in the tool box title line. Clicking the button activates the command. Your tool box may be displayed differently from Figure 2.1. The icon in the middle of the tool bar above the drawing area toggles the tool box display between floating, fixed at the left, invisible and fixed at the right of the window.

The command prompt area is located below the drawing area. This is where commands that you type appear along with the appropriate responses and prompts from AutoCAD.

The two little arrows at the lower left of the screen drawing area are AutoCAD's coordinate system icon or symbol. They point to the directions of the coordinate axes. In this case X is the horizontal axis and Y is the vertical and to locate a point four units to the right and three units up you would use the coordinates "4,3". (A third axis, Z, is available for 3D work and is perpendicular to the screen.) The "W" indicates that the WORLD or global coordinate system is active. This just means that when you give a pair of coordinates such as "4,3" that the location is calculated relative to the drawing's origin, which has the coordinates "0,0". You need not worry about the intricacies of coordinate systems; the default "X,Y,WORLD" is all that is required until Chapter 8. The coordinate system icon is on the screen as a reminder to the user, but it is not part of the actual drawing and so does not appear on plots.

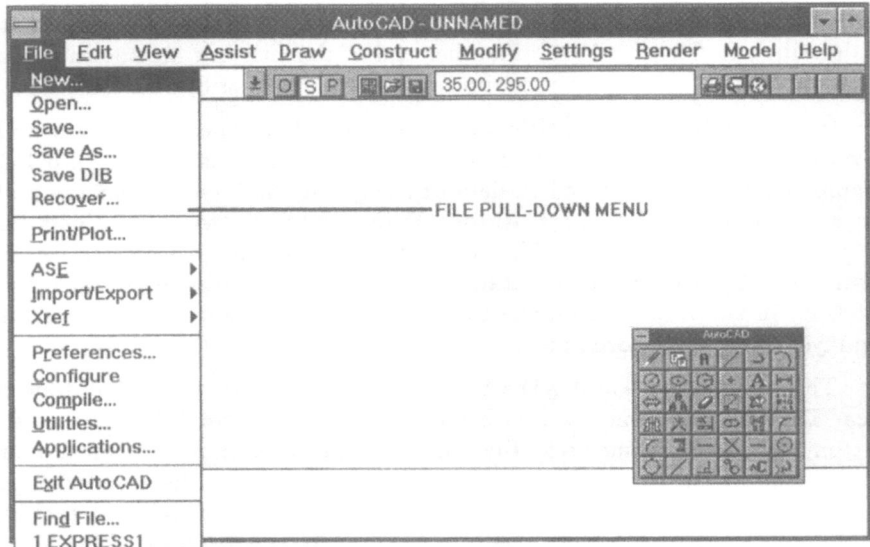

Figure 2.2 File pull-down menu

The File pull-down menu

Pull-down menus can be accessed by either of two methods. Firstly, using the generic Windows method, holding down the **ALT** key and pressing the underscored letter on the keyboard will bring up the appropriate menu. For example, pressing **ALT-F** causes the File menu to appear (Figure 2.2). To make it disappear without activating any of the commands press the **ESC** key. The second method is to move the cursor into the menu bar and click on the word **File**. As cursor is moved left or right on the menu bar the various key-words on the bar will be highlighted.

The File menu enables you to create a new drawing, open an existing drawing and save the current drawing to disk. Towards the bottom of the menu you will find the Exit AutoCAD which is used to end the drawing session. The file names given at the bottom of the menu will be the most recently edited drawings.

Drawing identification

With the **File** menu displayed pick the **NEW** option at the top. This causes the "Create New Drawing" dialogue box to pop up as shown in Figure 2.3.

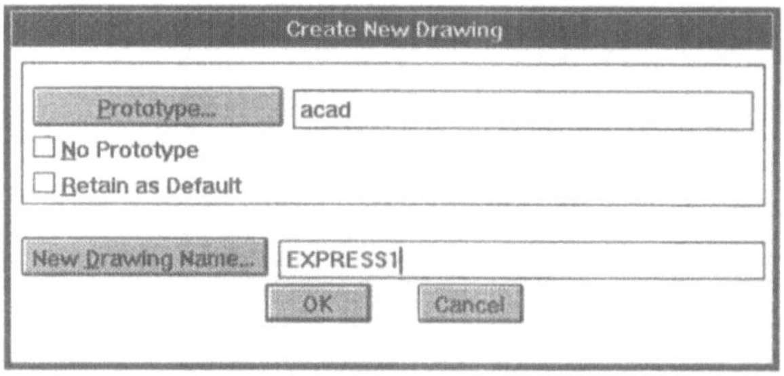

Figure 2.3 Create New Drawing dialogue box

Move the cursor to the input box beside "New Drawing Name" and click the left button once. Then type **EXPRESS1** and click the **OK** button.

The name can be from 1 to 8 characters long and can have letters or numbers. Lower case letters and upper case letters are fully interchangeable (a is equivalent to A etc.) but no full stops or blanks are allowed. AutoCAD appends the file extension ".DWG" onto the drawing name so that the full name of the file to be stored on the disk is EXPRESS1.DWG. With one exception, the user never specifies the ".DWG" part of the name when using AutoCAD. The exception in AutoCAD is when using the File/Utilities submenu for copying, deleting and renaming files. You would also need to specify the ".DWG" if you were using DOS on its own, say for disk tidying.

The drawing name can be preceded by a valid DOS pathname. For example by responding to the "New Drawing Name" prompt with a drawing name "C:\ACAD\EXPRESS1" you can specify that the drawing is to be stored on disk drive C: in the directory ACAD.

The top input box in Figure 2.3 contains the name "acad" as the prototype. The prototype is a drawing file that is copied into the new drawing. It will specify all the initial settings in AutoCAD for the new drawing.

Once you have named your drawing the title bar at the top of the window should change to the new name.

HAZARD WARNING! Never use the name ACAD for your new drawing as ACAD.DWG is used the default prototype used by AutoCAD. You can of course use a different drawing as a prototype. This is described in Chapter 3.

AutoCAD's menus

It is possible to type all of AutoCAD's commands using the keyboard. What you type appears on the command line and is executed when you hit

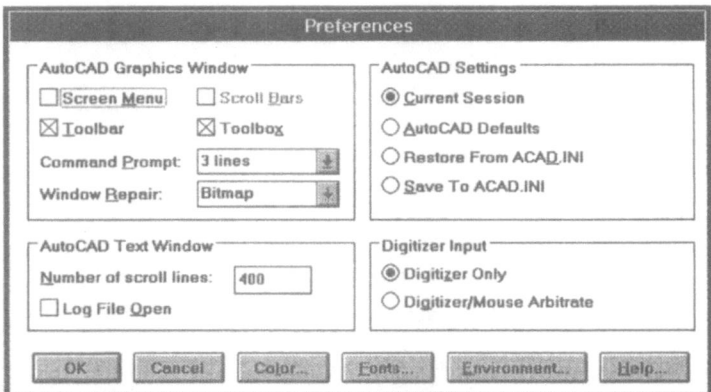

Figure 2.4 Preferences dialogue box

<ENTER>. However, to save time and typing errors and also to remind the user of the commands available AutoCAD has a system of menus and submenus. These provide convenient groups of similar commands. The tool buttons provide speedier access to the most frequently used commands. However, some of the icons on the buttons have pretty obscure representations of the commands.

Depending on which version of AutoCAD you are using there may be some slight differences between the menus displayed in this book and those that appear on your screen. A number of vendors supply AutoCAD with customized menus. The menus in *AutoCAD for Windows Express* are the "vanilla" files with no extra frills shipped with the English Version.

Menu preferences

If your AutoCAD window is missing some of the features mentioned above then it is possible that your system has non-standard defaults. The menus and tools that are displayed are controlled by the **Preferences** option on the **File** pull-down menu. Figure 2.4 shows the setup used for this book. The X's in the boxes for Toolbar and Toolbox indicate that these are active. To change any setting, move the cursor to the box and click once. Finally, click **OK** to confirm the new settings.

A note for experienced AutoCAD for Dos Users. If you are used to using AutoCAD's screen menus at the right-hand side of the graphics area then it is still possible to use them. Under the **File/Preferences** pull-down menu,

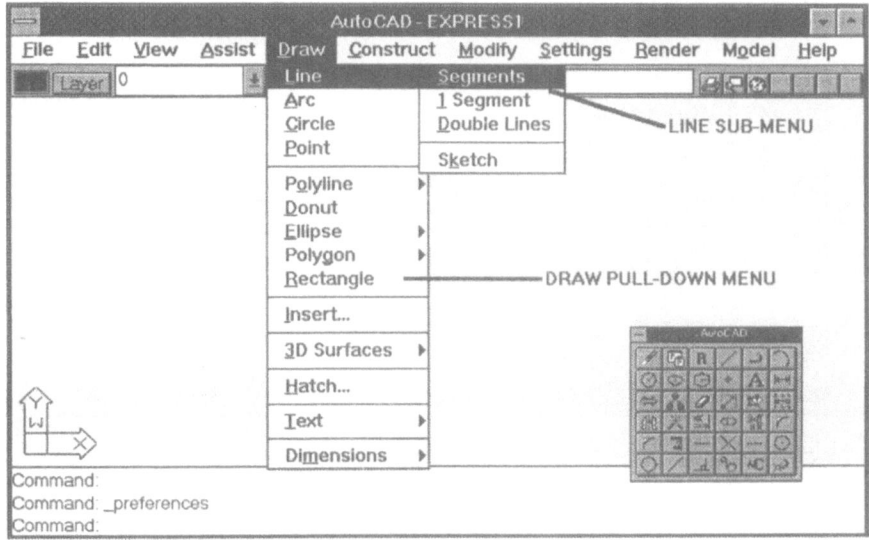

Figure 2.5 Line sub-menu

clicking the "Screen Menu" option so that an X appears in its box activates the old fashioned menu system. This is not recommended for new users and is likely to disappear in future AutoCAD releases.

How about a bit of doodling?

The menu bar at the top of the window consists of a cascading structure of menus and commands (Figure 2.5). At the end of each branch is a command. As an example, move the cursor so that the **Draw** lights up and press the pick button. The Draw pull-down menu appears. This contains a list of sub-menus and commands. The horizontal lines are purely to visually separate the different groups of commands. Sub-menus are indicated by the triangle symbol pointing right. Menu items with no triangle are executable commands.

If a column of icons appears instead of the names shown in Figure 2.5 move the cursor back to the menu bar and click **Settings** followed by **Menu Bitmaps** at the bottom of the subsequent menu. This toggles between the icons and text display of menus. The text menus are recommended for beginners. Now go back to the **Draw** pull-down menu.

Moving the cursor to **LINE** and picking causes the Line sub-menu to cascade as shown in Figure 2.5. Now pick **Segments** and draw a few joined

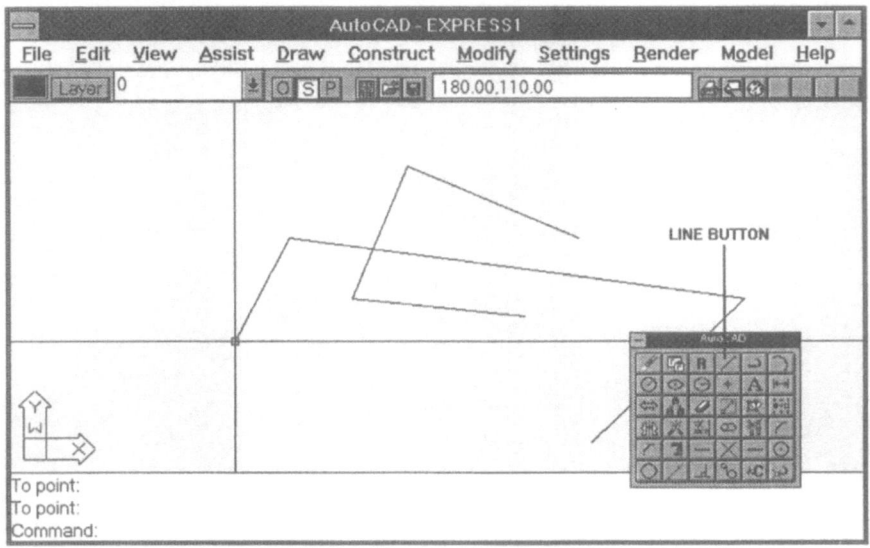

Figure 2.6 Doodles

lines. The sub-menu will disappear and at the bottom of the window the prompt line will display

Command: LINE From point:

and will wait for you to input the coordinates of one end. Move the cursor into the drawing area and press the pick button. This is taken as the start of the line and the prompt changes to

Command: LINE From point: To point:

waiting for the other end point to be input. Pick as many points as you want to create a series of connected line segments (Figure 2.6). To exit from the LINE command simply press <**ENTER**>.

This has been a quick squirrel hop along one part of the menu tree. Tasty morsels are also to be had along the other branches. As you use AutoCAD you will become familiar with the menus and the most useful commands. If you get lost on the menu tree, pressing **ESC** will return you to the graphics screen. Remember, commands can also be typed at the keyboard or picked from the floating tool box.

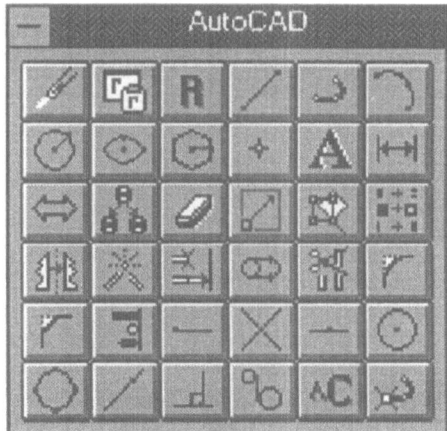

Figure 2.7 AutoCAD's box of tools

The tool box

One of the innovations in AutoCAD for Windows is the provision of the floating tool box (Figure 2.7). This contains 36 of the most frequently used commands. The icons are a bit obscure to start with but if you move the cursor slowly over them the name of the command appears at the top of the tool box. The LINE command can be executed by picking the fourth button from the left on the top row.

When the tool box is in its "floating" mode you can move it around the screen. Move the cursor into the title area of the tool box and hold down the left button on the mouse. Keeping the button pressed, drag the tool box to the desired location and release. As mentioned earlier, the icon in the middle of the tool bar above the drawing area can be used to toggle the display of the tool box. Click the icon four times to see the different options and return to the current setting.

Setting up the drawing environment

Now that you know a bit about communicating with AutoCAD, why not execute some commands and embark on some controlled CAD? The emphasis here is on your being in control and not accepting any old rubbish the computer might tempt you with in the name of "convenient defaults". The default drawing environment contains all the settings somebody somewhere found suitable for his or her own application. You can be sure that they were not meant

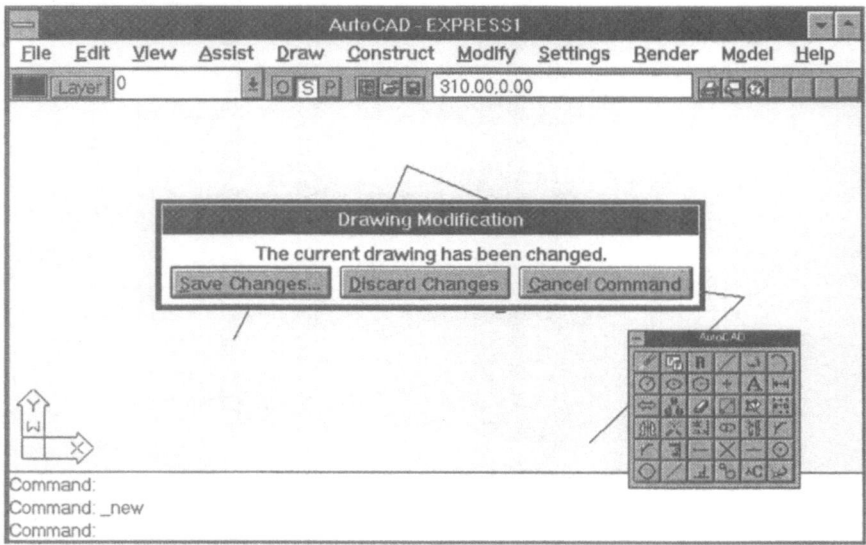

Figure 2.8 Dumping an unwanted drawing

for you. Even if it is acceptable for one drawing the environment may need changes for the next.

For this section you will need a clear drawing, so if you have already drawn some doodle lines, follow the procedure below to discard the current drawing. If your drawing is empty, then skip to the "Drawing size" paragraph.

Pick **File** from the menu bar at the top of the window followed by **New**. The dialogue box in Figure 2.8 will appear to warn you that the drawing has not been saved. You now have the option to save or discard the changes to the drawing file or to cancel the operation and return to the drawing editor. In this case pick **Discard Changes**.

This will discard the current drawing and then display the "Create New Drawing" dialogue box. The drawing name should still be displayed as "EX-PRESS1". Click on the **OK** button to accept this name for the new drawing.

This starts a new drawing, again called EXPRESS1.DWG. If the drawing already exists you will get a warning message asking if you want to overwrite it. Type "**Y**" to overwrite.

Drawing size

All AutoCAD drawings are made to *full scale*. One of the principal reasons for
this is so that when you use AutoCAD's automatic dimensioning it will give
the correct lengths rather than scaled ones. Thus the drawing size will depend
on the size of the items being drawn. It will also depend on the working units.
For example, to draw the architectural layout of an office measuring 23.8m
by 15m you would choose a size big enough to contain the whole floor, plus
a bit to spare for extra views. A good drawing size in this case might be 29.5
units wide by 21 units high. If you were using millimetres as the base unit
rather than metres, then the drawing size would be 29,500 by 21,000. (Note
that this is approximately the same ratio as an A4 sheet and other sizes in the
A series.) This facilitates the layout of the drawing for subsequent plotting.
For the drawing EXPRESS1 you will require it to be 65 units by 45 units.
The LIMITS command allows us to do this. Pick **Settings** from the menu
bar and then **Drawing Limits**. This executes the LIMITS command and the
command line area should echo the following. What you have to type is in
bold text.

> Command: LIMITS
> Reset model space limits:
> ON/OFF<Lower left corner> <0.00,0.00>: **<ENTER>**
> Upper right corner <420.00,297.00>: **65,45 <ENTER>**
> Command: **LIMITS <ENTER>**
> Reset model space limits:
> ON/OFF<Lower left corner> <0.00,0.00>: **ON <ENTER>**
> Command:

Pressing **<ENTER>** in response to the lower left prompt means that you
accept the default setting of positioning that corner at the WORLD origin,
0,0. AutoCAD always displays the default values between angular brackets,
< > and any time you wish to accept this value simply press **<ENTER>**. If
you wish to override the default then key in the desired values as in the upper
right prompt above. If the defaults offered by your computer are different from
420.00,297.00, don't worry. Simply replace whatever the values are by **65,45
<ENTER>**. The second execution of LIMITS is to turn the limit checking
facility on. This prevents anything being drawn outside the limits by mistake.
This is especially important when using AutoCAD for Windows since the
shape and size of the drawing window usually does not match the actual limit
proportions.

At this point if you move the cursor around the drawing area you will
notice that the coordinate read-out in the tool bar or status line is giving
similar values as before. So, even though you have changed the drawing size

the display shows the old size. To display the whole of the new drawing size type:

Command: **ZOOM** <**ENTER**>
All/Center/.../<Scale(X/XP)>: **A**

The response to the type of zoom required can be truncated to whatever Auto-CAD displays in CAPITAL letters, in this case "A" for "All". This command can be found by picking **View** from the menu bar, then **Zoom** followed by **All**. It's a toss up whether it is quicker to type or to pick from the menus. This command works like a zoom lens in a camera allowing magnification and demagnification of the image. ZOOM is fully described in Chapter 3. Now, if you move the cursor around, the coordinates will reflect the current drawing size.

While it is important to consider the drawing size before you embark on an AutoCAD drafting session, it is not essential. You can alter the drawing size at any time during editing, but it's more efficient to get it right first time!

If the cursor now appears a bit jumpy when moved about the drawing area you can adjust the snap increment. The SNAP value, when enabled, limits cursor movements to discrete steps. If the steps are too large the cursor will seem jumpy. If it is too small or switched off then it can be difficult to accurately pick points in the drawing. The following short command sequence sets a snap value of 1 unit which is about right for the subsequent exercise. In this instance the command is typed at the keyboard. Further examples of SNAP are given in Chapter 3.

Command: **SNAP** <**ENTER**>
Snap spacing or ON/OFF/Aspect/Rotate/Style <>: **1**

The next choice you have for the drawing environment is the system of units to be used. The standard AutoCAD default settings use decimal units and decimal angles. If the coordinates on the status line are in decimal format then you can skip to the next section, on "Drawing lines". If it is in feet and inches, or you are not sure, then use AutoCAD's UNITS command. This is on the Settings pull-down. Pick **Settings** from the menu bar and then **Units Control**. The dialogue box shown in Figure 2.9 should appear. Pick **Decimal** in the Units column and **Decimal Degrees** for angles.

To adjust the precision of the coordinate readout click on the Windows pull-down icon (down pointing arrow) to the right of the box below the word "Precision". This brings the pull-down menu with varying numbers of zeros after the decimal point. Use the scroll arrows on the right to move up or down. When **0.0000** is highlighted, pick it with the cursor. Repeat this procedure for the an angular precision of **0.00**.

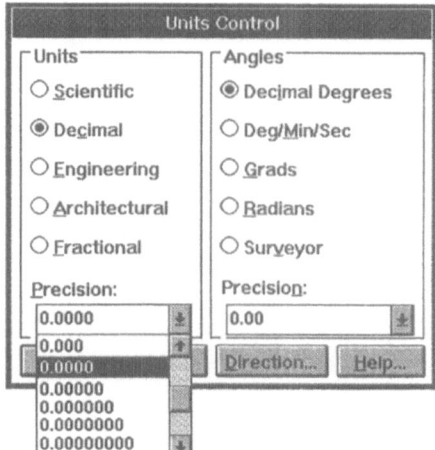

Figure 2.9 Units control settings

Finally, with angles you can choose which direction is represented by 0 degrees. Pick **Direction** from the bottom of the dialogue box to make sure that East is 0 degrees. Click **OK** once to close the Direction Control box and once more to close the Units Control box. This procedure should ensure that the examples in this book match the displays on your screen.

There are many other drawing "environment" settings that can be specified but for the time being we will leave it at Limits and Units. Other settings will be introduced when they become useful.

Drawing lines

Having done all that work on setting up the drawing environment, you are in a position to do some controlled drafting. The workhorse of any drawing is the humble line. Indeed, ultimately, every drawing entity can be reduced to a series of straight lines (curves consist of a very large number of tiny straight lines).

The AutoCAD LINE command allows you to construct any number of independent line segments. To draw a square 15 x 15 pick **Draw/Line/ Segments** in that order from the menu bar and type the coordinates given below. In response to the final "To point:" prompt type the word "close".

Command: LINE
From point: **5,5 <ENTER>**   (A)

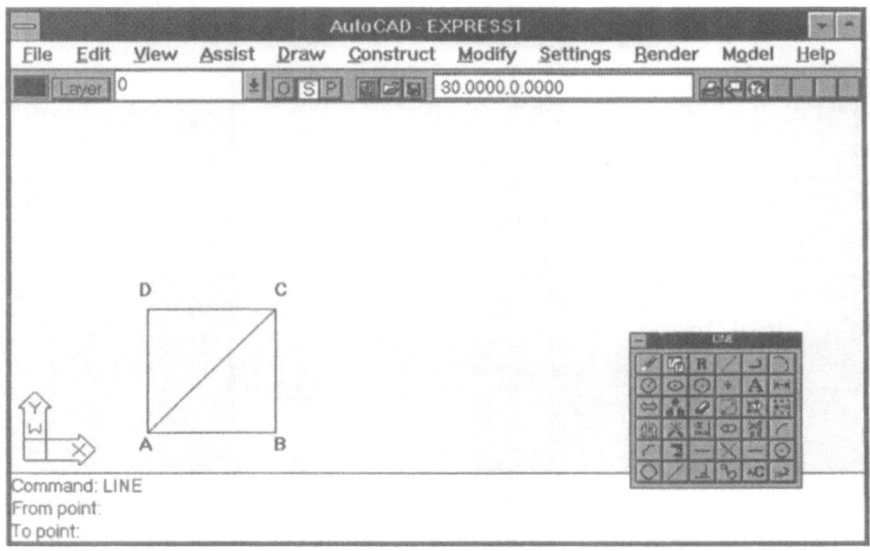

Figure 2.10 Drawing a square

> To point: **20,5 <ENTER>**   (B)
> To point: **20,20 <ENTER>**   (C)
> To point: **5,20 <ENTER>**   (D)
> To point: **CLOSE <ENTER>**

AutoCAD's first response to the LINE command is to ask for a start point and then the end point. The program assumes that you want to continue drawing lines until you tell it otherwise. By typing "**CLOSE <ENTER>**" you cause AutoCAD to join the end point of the last line segment to the very first point input. Now draw a line along one of the diagonals of the square from A to C in Figure 2.10. These letters won't appear in your drawing, they are just used to clarify this text.

> Command: **LINE <ENTER>**
> From point: **5,5 <ENTER>**   (A)
> To point:

If you move the cursor around the drawing area the line "rubber bands" and extends from the point 5,5 to wherever the cursor is located. As the line rubber-bands the coordinate display on the tool bar may jump to polar notation ie distance and angle. If this happens press ˆ**D** (CTRL+D) to toggle it back to x,y coordinates.

You can now pick the point at the other end of the diagonal or type the coordinates.

> To point: **20,20** <**ENTER**>   (C)
> To point: <**ENTER**>

As mentioned above, AutoCAD keeps replying with the prompt "To point:". When you have finished inputting lines you can exit the LINE command by pressing <**ENTER**> without giving any point. Alternatively, you can hit the space bar on the keyboard instead of the ENTER key. In most AutoCAD commands it is possible to interchange use of the space bar and the ENTER key. As the former is much larger it is much easier to locate (or harder to miss!). A third way to get out of the LINE command (and indeed any other AutoCAD command) is to CANCEL it. Press ^C (CTRL and C on the keyboard) to cancel the current operation and return to the "Command:" prompt. The ^C icon can also be found second from right on the bottom row of the tool box. Remember ^C as it can get you out of trouble when things go wrong. CANCEL also appears in dialogue boxes, ready to come to the rescue.

What if you want to draw two lines which are not connected? Use **Draw/Line/1 Segment**. AutoCAD gives some assistance by using the same LINE command as used above but adds an <**ENTER**> after the second point has been picked. To draw the two parallel lines, EF and GH, in Figure 2.11 pick **Draw** from the menu bar, then **Line** followed by **1 Segment**. Then pick the point 5,25 and pick or type the second point 20,30. AutoCAD automatically exits LINE and returns to the command prompt.

> Command: LINE From point: **pick 5,25**   (E)
> To point: **20,30** <**SPACE BAR**>   (F)
> To point:

Now pick the **Draw/Line/1 Segment** sequence once more.

> Command: LINE From point: **5,30** <**ENTER**>   (G)
> To point: **20,35** <**ENTER**>   (H)
> To point:
> Command:

AutoCAD's command memory

AutoCAD remembers the last command executed and it offers that command again if you press <**ENTER**> or <**SPACE BAR**> at the "Command:" prompt without typing anything else. The right-hand mouse button may also have this function. This is an obvious time saver when doing repetitious sequences like drawing lots of lines.

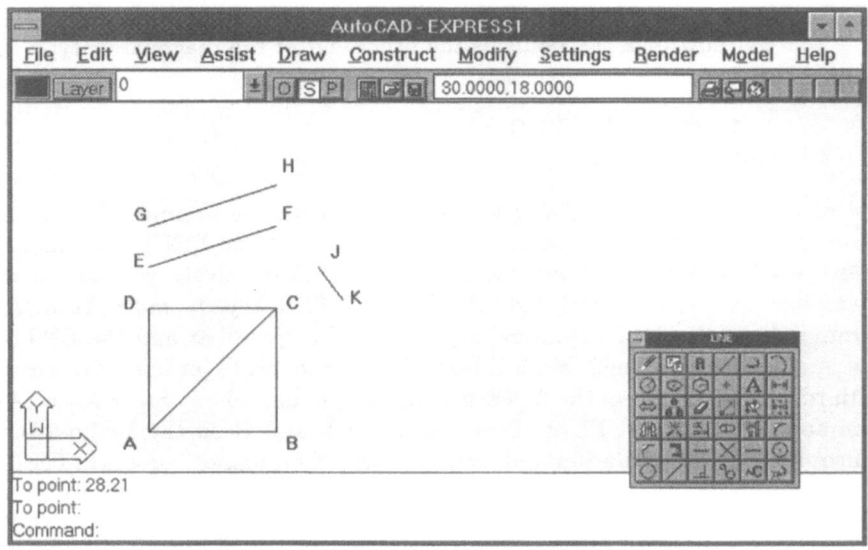

Figure 2.11 Adding some lines

To draw the line JK in Figure 2.11 press <**ENTER**> and pick the points 25,25 and 28,21 and press <**ENTER**> to get back to the Command: prompt.

```
Command: <ENTER>
LINE From point: 25,25                                    (J)
To point: 28,21                                           (K)
To point: <ENTER>
Command:
```

We have already covered the "close" option. It makes a closed polygon of the lines by joining the most recently input point to the first point (ie. the "From point:") of the sequence. Close, which can be abbreviated to "c", will only work if more than one line has been drawn in the current LINE operation. The "current LINE operation" means all the inputs from ONE pick issuing of "LINE" at the Command prompt.

Another sub-command that only works within the current LINE operation is the undo option. If, while drawing a sequence of connected line segments, an incorrect "To point:" is picked, you can immediately undo the last pick and re-input a new point. To illustrate, pick the **Draw/Line/Segments** command

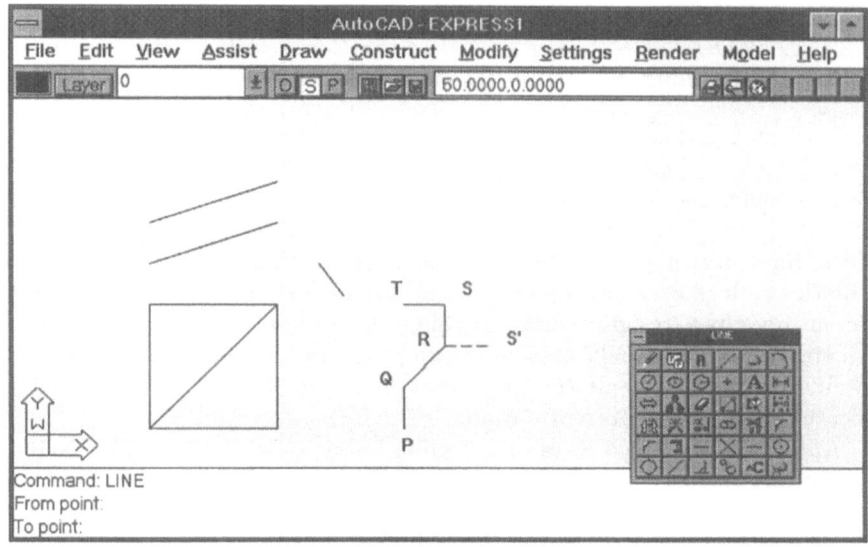

Figure 2.12 The undoing of a line

and then the points P, Q, R and S (see Figure 2.12). Without leaving the LINE command type **undo** at the "To point:" prompt. The point S disappears and you are prompted for a new "To point:" and can pick the points S' and T.

Command: **LINE** <ENTER> From point: **35,5** <ENTER>	(P)
To point: **35,10** <ENTER>	(Q)
To point: **40,15** <ENTER>	(R)
To point: **45,15** <ENTER>	(S')
To point: **undo** <ENTER>	
To point: **40,20** <ENTER>	(S)
To point: **35,20** <ENTER>	(T)

You can undo all the line segments back to the point, P, in this manner but only if they are all part of the same LINE operation.

There is another UNDO command in AutoCAD which can reverse any previous command. The one within the LINE command only works for back-tracking the line segments. If you type UNDO at the "Command:" prompt it will back-track the previous command or commands. This is described further in Chapter 4.

HAZARD WARNING! Typing UNDO at the Command: prompt can be the undoing of everything. If you undo too much, use the REDO command immediately.

Linetypes and scales and colors

All the lines drawn so far have been standard continuous ones, either white on black or black on a white background. Engineers and architects use many different linetypes to signify different things. Center lines, hidden lines, dashed lines etc are all commonly used in drawings. As you would expect AutoCAD contains definitions of all the important linetypes. Colors are also used to indicate certain types of information.

AutoCAD gives two ways of assigning linetypes and colors. The method given in this exercise is the "quick and nasty" method. We just tell AutoCAD to change the linetype and color for subsequent entities (lines, circles, text etc). The second method, described in Chapter 3, is to divide the information on the drawing into a series of "layers". For example, the plans of a house might contain a layer of plumbing information, a layer of electrical layout and layers for walls etc. Each layer is then assigned an appropriate default color and linetype. You then have to tell AutoCAD which layer you wish to work on.

Getting back to the quick and nasty method, take a closer look at the tool bar at the top of the drawing area (Figure 2.13). In the current drawing, EXPRESS1, there is only one layer which is called "0". This name appears in the tool bar after the word "Layer". Just to the left of the "Layer" button is a box which indicates the current color. Pick **Settings** followed by **Entity modes** from the menu bar. Alternatively, moving the cursor into the "color box" and clicking also causes the Entity Creation Modes dialogue box to appear.

The initial color is called "BYLAYER", ie whatever the layer has been set to. The default layer color is white on black (or black on white depending how your screen is configured). Pick **Color** from the dialogue box and then pick your favorite shade of blue from the subsequent palette that appears and pick **OK**. Then pick **Linetype** from the dialogue box followed by **Dashed**.

If Dashed doesn't appear on your list try picking **Next** to get another page of linetypes. If there is only one linetype available, namely, continuous you will have to load the others. Click **Cancel** and type the following sequence.

Command: **LINETYPE**
?/Create/Load/Set: **LOAD**
Linetype(s) to load: **DASHED,CENTER**
File to search <acad>: <**ENTER**>

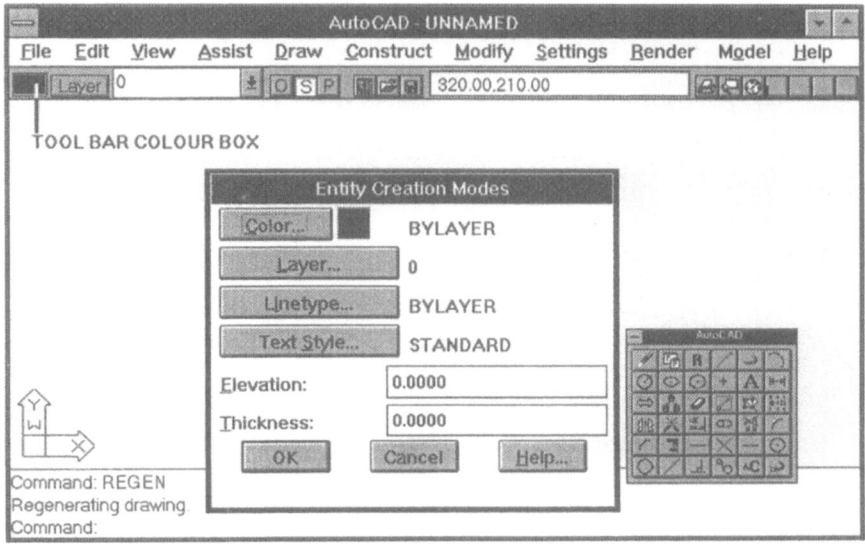

Figure 2.13 Toolbar and Entity Creation Modes

You may be prompted for this file via a dialogue box (Figure 2.14). If so, click the cursor on the file, ACAD, followed by "OK". Then try **Settings/Entity modes** again.

If the dialogue box shown in Figure 2.14 did not appear, you can make it do so by changing the setting of AutoCAD's "FILEDIA" variable. Set the value to 1. A value of 0 disables the dialogue box.

    Command: **FILEDIA <ENTER>**
    New value for FILEDIA <0>: **1 <ENTER>**

Now draw the lines U to V to W to X to close as shown in Figure 2.15.

    Command: **LINE <ENTER>**
    From point: **60,40 <ENTER>**   (U)
    To point: **35,40 <ENTER>**   (V)
    To point: **35,30 <ENTER>**   (W)
    To point: **60,30 <ENTER>**   (X)
    To point: **C <ENTER>**

Did the lines come up dashed? If not, this may be because the dashes are too small or too big. You can control the size of the dashes by altering the linetype scale, or as AutoCAD calls it, the LTSCALE. Initially, the LTSCALE

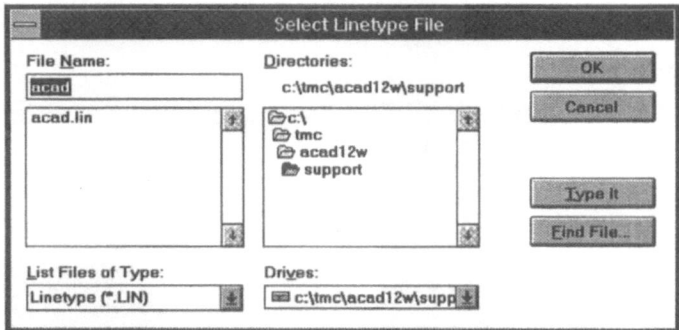

Figure 2.14 Select linetype file dialogue box

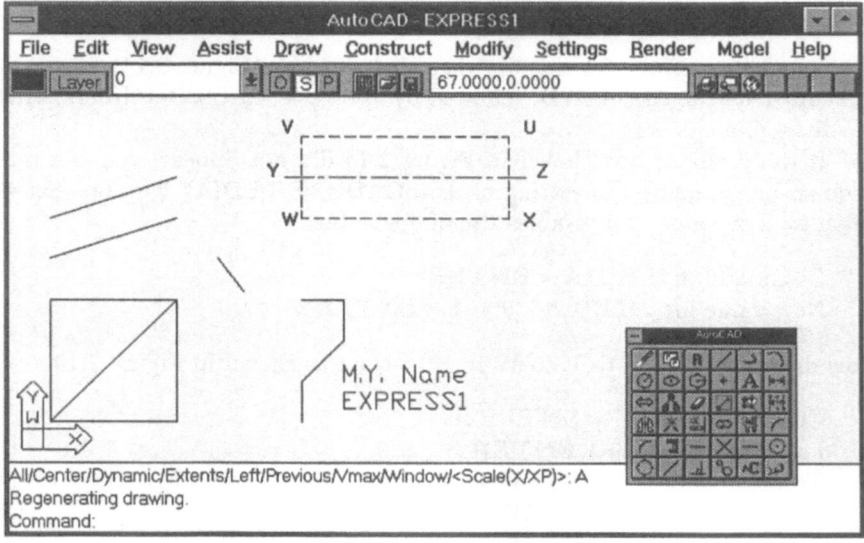

Figure 2.15 Dashed lines

has a value of 1. For the DASHED linetype this means that you have one
dash and one space per drawing unit. For consistency with Figure 2.15 set the
LTSCALE to 0.1.

> Command: **LTSCALE: <ENTER>**
> New scale factor <1.0>: **0.1 <ENTER>**
> Command: **REGEN <ENTER>**   (This may happen automatically.)
> Command:

The REGEN command forces AutoCAD to re-calculate all the lines using this
new linetype scale. Depending on your default settings this regeneration may
occur automatically after LTSCALE. The larger the drawing limits the larger
LTSCALE so that the dashes are visible.

Dashed and center are only two of the linetypes supplied with AutoCAD.
To find out what others are available type the command, **LINETYPE.**

> Command: **LINETYPE <ENTER>**
> ?/Create/Load/Set: **? <ENTER>**
> File to list <ACAD>: **<ENTER>** or Click **OK** from dialogue box

This should display the various types as shown in Table 2.1. Press **<ENTER>**
or RETURN for the rest of the list.

In Windows an AutoCAD Text window is opened to show the linetypes.
This can be closed by picking the Control menu button in the top left corner
followed by close or by the hot key combination **ALT+F4**. AutoCAD for DOS
users can get back to the graphics screen by pressing the F1 key.

This list contains both the American and English spelling of CENTER.
Either spelling is allowed. Only linetypes that have been loaded into the current
drawing will appear in the Entity Modes/Linetype dialogue box.

As an alternative to using the dialogue box you can change linetypes by
using the "Set" option of LINETYPE. For example, the following commands
would be used to draw a center-line through the last rectangle without changing layer.

> Command: **LINETYPE <ENTER>**
> ?/Create/Load/Set: **S <ENTER>**
> New entity linetype (or ?) <DASHED>: **CENTER <ENTER>**
> ?/Create/Load/Set: **<ENTER>**
> Command: **LINE <ENTER>**
> From point: **33,35 <ENTER>**   (Y)
> To point: **62,35 <ENTER>**   (Z)
> To point: **<ENTER>**
> Command: **LINETYPE <ENTER>**
> ?/Create/Load/Set: **S <ENTER>**

Table 2.1 Standard linetypes

Name	Description
BORDER	-- -- · -- -- · ---- · -- -- · -- --
BORDER2	--·--·--·--·--·--·--·--·--·--·--·--
BORDERX2	---- ---- · ---- ---- · ---- --
CENTER	---- - ---- - ---- - ---- - ---- - ---
CENTER2	--- - --- - --- - --- - --- - --- - --
CENTERX2	--------- -- --------- --- --------- -
CENTRE	---- - ---- - ---- - ---- - ---- - ---
CENTRE2	--- - --- - --- - --- - --- - --- - --
CENTREX2	--------- -- --------- --- --------- -
DASHDOT	-- · -- · -- · -- · -- · -- · -- · --
DASHDOT2	--·--·--·--·--·--·--·--·--·--·--·--·
DASHDOTX2	----- · ----- · ----- · ----- ·
DASHED	-- -- -- -- -- -- -- -- -- -- -- --
DASHED2	- - - - - - - - - - - - - - - - - -
DASHEDX2	----- ----- ----- ----- ----- ---

– Press Return for more –

New entity linetype (or ?) <CENTER>: **BYLAYER <ENTER>**
?/Create/Load/Set: **<ENTER>**

The second execution of LINETYPE was to reset it to the individual default
settings of the layers.

While this facility gives flexibility to your drafting, it must be used sparingly and with caution. Since different types of line usually convey different
types of information it makes sense to collect them onto individual layers as
described in the next chapter. If at all possible use only one linetype per layer.

Similarly, when we are finished with the special colors it is advisable to
reset it to BYLAYER. This can be done by using the COLOR command or
via the dialogue box (the English spelling of colour is also recognized).

Command: **COLOR <ENTER>**
New entity color <BLUE>: **BYLAYER <ENTER>**

Saving the drawing

While working in AutoCAD you can periodically save the drawing and any
changes you have made to it. To save EXPRESS1 pick **File** from the menu bar

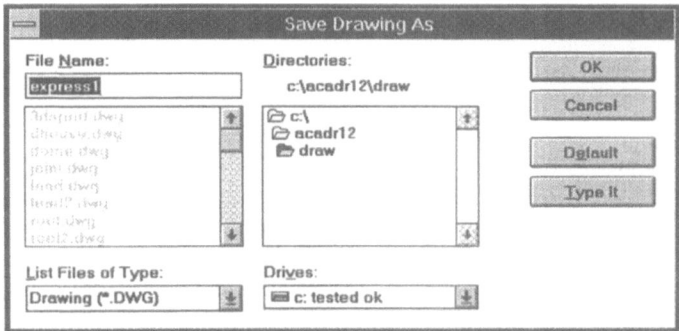

Figure 2.16 Save Drawing dialogue box

followed by **Save....** Alternatively type **SAVE <ENTER>** at the Command prompt. This displays the "Save Drawing As" dialogue box (Figure 2.16). Click **OK** to accept the default drawing name, which in this case is "EXPRESS1"

> Command: **SAVE <ENTER>**

You should use SAVE approximately every 15 to 20 minutes during a drawing session. Always do a SAVE before attempting difficult or large tasks (eg copying a large layout) or before attempting something new. Some commands are categorized in this book as potentially dangerous (HATCH, HIDE etc) and a SAVE should be done before executing them.

If the file EXPRESS1.DWG already existed on the disk you would get a warning message from AutoCAD. If you then proceed with the SAVE, the EXPRESS1.DWG file will be copied to EXPRESS1.BAK and a new .DWG file made. This backup file will be stored in the same directory as the original .DWG file.

There is an autosave option in AutoCAD's configuration. This is described in Appendix A and is for more experienced users.

Adding some text

Before you leave EXPRESS1 as a completed drawing, you can put your name to it, as shown in Figure 2.15. You can insert your own name in place of "M.Y. Name". Pick **Draw**, then **Text** and **Dtext** from the menu bar.

> Command: **DTEXT <ENTER>**
> Justify/Style <Start point>: **40,10 <ENTER>**
> Height <3.00>: **2 <ENTER>**

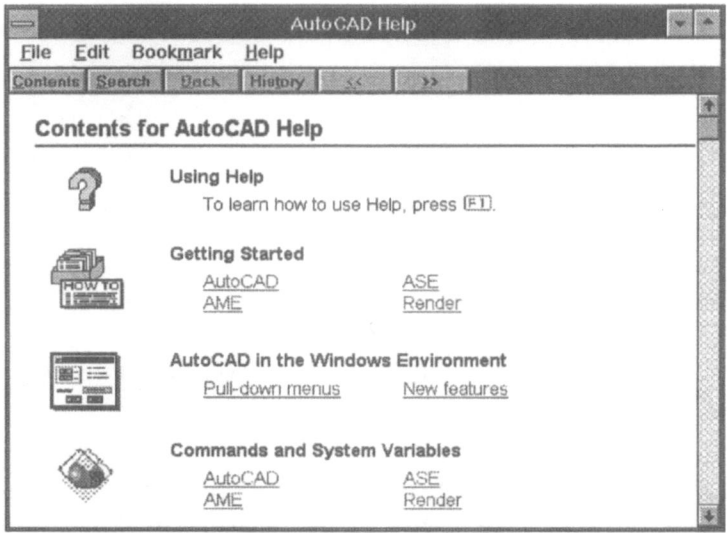

Figure 2.17 AutoCAD Help contents

Rotation angle <0>: **<ENTER>**
Text: **M.Y. Name <ENTER>**
Text: **EXPRESS1 <ENTER>**
Text: **<ENTER>**

HELP!

That piece of text completes the drawing part of this chapter. Before finishing this editing session it's worth taking a quick look at AutoCAD for Windows HELP facility. This provides information about all of AutoCAD's commands, system variables and all the Windows specific features of the program. It is structured in a similar manner to Help facilities for other Windows applications.

As with Windows the AutoCAD help facility can be activated by pressing the **F1** key. It can also be found at the right hand end of the menu bar. From the menu bar pick **Help** followed by **Contents**. The AutoCAD Help window opens as shown in Figure 2.17. Clicking on any of the underlined items will cause a further page of contents about that item to appear. When help extends over one screen then the scroll bar at the right can be used.

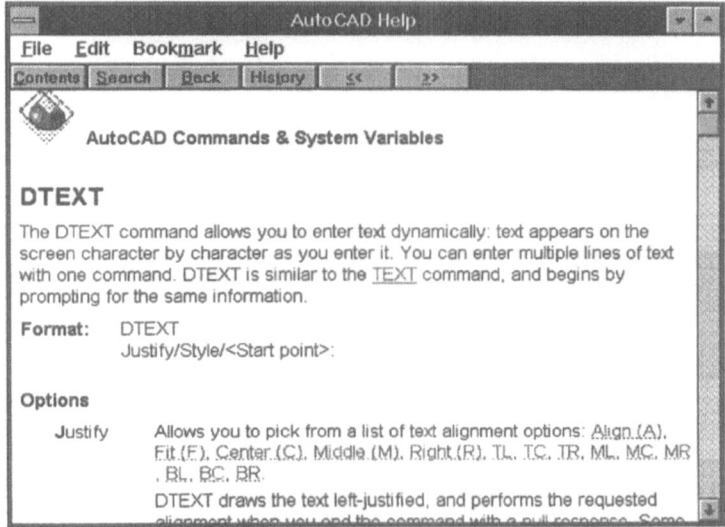

Figure 2.18 DTEXT help

You can find yourself moving down through the levels of help quite quickly. If you get lost, the "Contents" button will bring you back to Figure 2.17. The "Back" button steps back one screen at a time while "History" will give you a list of all the screens that have been accessed. Pressing F1 again at the contents screen gives you help about the Help facilities. You can exit HELP and get back to the drawing screen by picking **File** and **Exit** or by double clicking the Control Menu box in the top left corner of the Help Window.

By way of an example pick **AutoCAD** from the Getting Started section. This gives a screen-full of new topics. Now pick **DTEXT**. This should tell you all about that command (Figure 2.18). Related topics appear as green text with solid underline. Click on these words for information about them. Items with dotted underlined green text lead to pop-up definitions. Click the mouse anywhere to make the definition disappear.

The "Search" button is a quick way of finding help on a specific command or topic. Clicking the **Search** button gives another dialogue box (Figure 2.19). Move to the keyword box and type **DTEXT**. The keywords list then shows all the items relating to DTEXT. Pick one of this list and then pick the **Go to** button. Pick the Close button if you do not want to proceed. Remember to exit from Help, pick **File** and **Exit** from the AutoCAD Help Window.

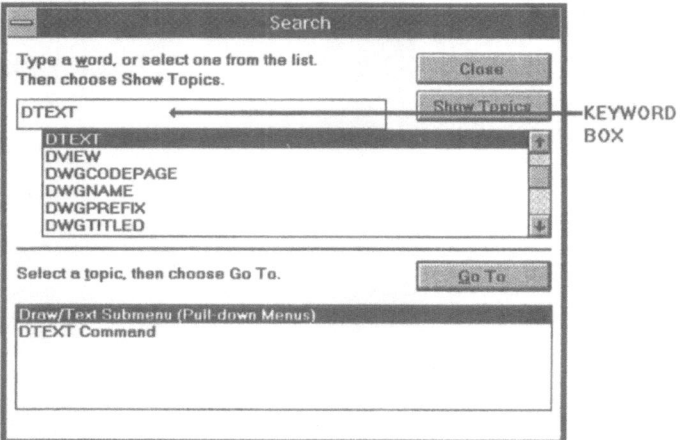

Figure 2.19 Search for help

Help - DOS Extender Version

To get help in the DOS version of AutoCAD type **HELP** at the Command prompt or pick it from the menu bar. A dialogue box will appear with all the AutoCAD commands listed. Enter the name of the command in the "Help Item" field at the bottom of the dialogue box. There is also a help index button for searching the help file. Pick the command name from the index and then pick OK.

Context sensitive Help

In both the DOS and Windows versions you can get help in the middle of an operation. The information displayed will relate to the command in progress. This is done by typing **'HELP** or **'?** at the command line. Be sure to include the single quote before help. For example, start the LINE command and then ask for help.

Command: **LINE**
From point: **'HELP**

This will tell you all about the line command. You can exit help as before and the LINE command will resume.

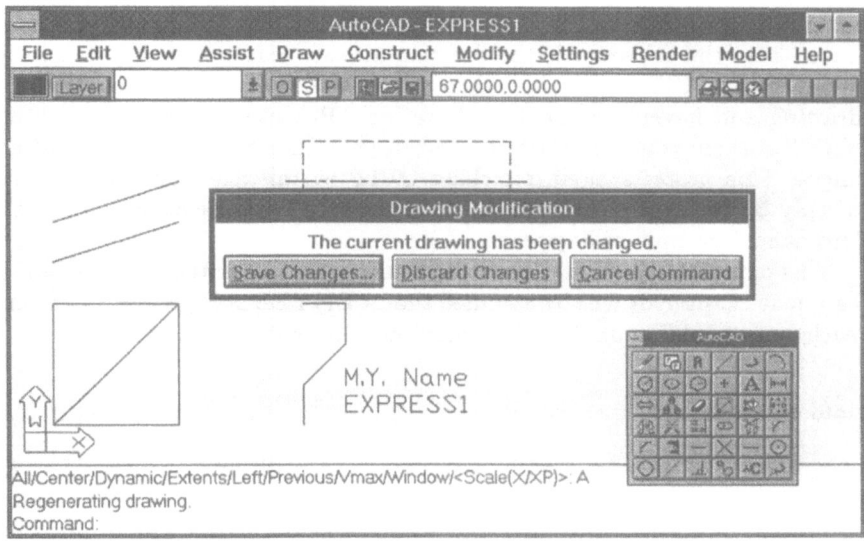

Figure 2.20 Drawing modification dialogue box again

Finishing up

To finish a drawing editing session you can either start editing another drawing or exit AutoCAD altogether. To switch to a new drawing pick **File/New**, to edit an existing drawing pick **File/Open** or to leave AutoCAD pick **File/Exit AutoCAD**. As we have not yet saved the last changes the "Drawing Modification" dialogue box will pop up again as shown in Figure 2.20.

Pick **Save Changes...** and pick **OK** in the file dialogue box to accept the default file name, EXPRESS1. This will make a copy of the drawing in its current form to the file called "EXPRESS1.DWG" and if you have picked "Exit AutoCAD" will return you to Windows.

NB: Keep a safe copy of this drawing file as it will be useful for the exercise in Chapter 4.

If you did not want to keep the new drawing (or changes to an old one) you can leave the editor by using the Discard Changes option. You can also do this by issuing the QUIT command.

Command: QUIT <ENTER>

Really want to discard all changes to drawing? YES <ENTER>

This provides a quick exit from AutoCAD's editor and is useful if you don't want to keep the drawing. If the drawing already exists then it is left in its original form. QUIT is suitable when you have only been scanning through a drawing and haven't made any alterations. Because of the serious nature of QUIT's usage you are asked to confirm your intentions to discard all the changes. This protects against picking QUIT by mistake from the menu. If you reply by typing NO <ENTER> then the QUIT will be cancelled and the "Command:" prompt will appear.

This completes your first non-stop journey on the AutoCAD Express. In subsequent chapters it will be assumed that <**ENTER**> is pressed at the end of each command line or that the commands are picked from menus.

A note on file security

Many things can go wrong with computers and disks and occasionally with AutoCAD itself. The words "FATAL ERROR" have an ominous ring and it is very annoying when such an event occurs. The only sane way to approach the inevitable is to prepare well, so that when it happens the consequences are minimized. Don't be depressed by all this doom and gloom. With adequate preparation the impact of a computer crash is cushioned.

It is good practice to have more than one copy of your work. AutoCAD provides facilities for this in two ways. Firstly, if you are going to edit an existing drawing file, then AutoCAD automatically makes a copy of the drawing. If the original was called "EXPRESS1.DWG" then the backup copy is called "EXPRESS1.BAK". This is certainly a very useful security measure but in itself is not fool proof. You should use the Windows File Manager or DOS to make further copies onto other disks or onto a tape streamer. With all important computer files it is recommended to have at least two copies on your working disk, and one safe copy on another disk which should be kept in a safe place. Regular backing up of files is *essential.*

Summary

This chapter has introduced many of AutoCAD's facilities and its methodology. In this way the chapter provides the basis for understanding the way all the commands work.

You should now be able to:

Load the AutoCAD program.
Create a new drawing file.
Pick commands from pull-down menus and the tool box.
Explore the pull-down menus.

Set up suitable AutoCAD drawing limits.
Draw lines and squares.
Undo lines.
Use dialogue boxes to select files.
Load extra linetypes.
Assign linetypes and colors to them.
Alter the linetype scale.
Save a drawing file every 15 minutes.
Make extra backup files.
Navigate the AutoCAD Help Facility
Quit a drawing without saving.
End a drawing, saving all changes.
Exit from AutoCAD.

Chapter 3 CURSOR AND DISPLAY CONTROL

General

Drawing with a mouse or tablet cursor is analogous to drawing freehand with a pencil. In engineering drawing we rely on a host of drawing aids and equipment. T-squares, rulers, setsquares and compass are but a few of the tools necessary to control the positions and sizes of drawn objects.

Similarly, we need electronic protractors and digital dividers when using AutoCAD. This chapter covers some of the useful cursor control facilities within AutoCAD. Using these facilities will make your drawings into exact graphical representations. You will learn how to set up construction lines. These will then be used to reference drawing items to the key construction points for making a replica of Paris's Eiffel Tower. The half tower shown in Figure 3.1 is the goal of the current exercise. The tower will be completed in Chapter 5.

For this chapter you will create a new drawing file, called EXPRESS2. The first task is to set up the drawing environment. Using the same limits as before, (0,0) to (65,45) you will then load a number of AutoCAD's linetypes. As this drawing will contain a number of different categories of information you will be introduced to AutoCAD's layering facility.

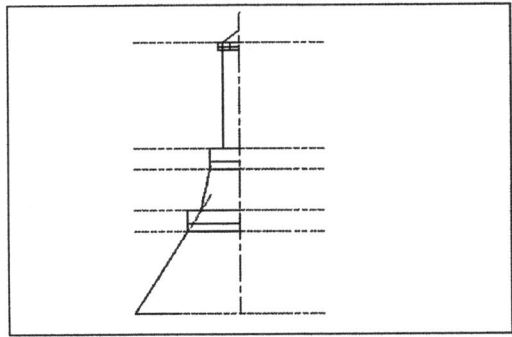

Figure 3.1 The Half full tower

Start up AutoCAD for Windows as described in Chapter 2. Set up the drawing environment using LIMITS and load the LINETYPEs.

Command: **LIMITS**
Reset Model space limits:
ON/OFF/<Lower left corner> <0.00,0.00>:
Upper right corner <420.00,297.00>: **65,45**
Command: **LIMITS**
Reset Model space limits:
ON/OFF/<Lower left corner> <0.00,0.00>: **ON**
Command: **ZOOM**
All/Centre/.../<Scale(X/XP)>:**A**

Remember turning the limits on stops things being drawn outside the drawing area. The Zoom All command resizes the viewing area to the new limits.

Command: **LINETYPE**
?/Create/Load/Set: **LOAD**
Linetype(s) to load: **CENTER,CENTRE,DASHED,HIDDEN**
File to search <acad>: **<ENTER>**

You will be prompted to input the file via the dialogue box as in the last chapter (Figure 2.14). Click the file, ACAD, and then OK. You will only really need the centre-line definition for this exercise but you might like to experiment with the others. You may get a message to say that some or all of the linetypes have already been loaded. Just type "Y" to reload it. You can chose an appropriate scale for the dashed and centre-lines.

Command: **LTSCALE**
New scale factor <1.0>: **0.1**

Use the same value for LTSCALE as you used in EXPRESS1.DWG. Finally, make sure that the drawing units are decimal (see Chapter 2). This reduction in the linetype scale is necessary to give the dashes sensible proportions within the new drawing limits.

Layers

Having decided on the size, it is now worth considering how your drawing is to be organised. Using AutoCAD's LAYER facility is the most efficient way of doing this. The layers of a drawing can be considered as a series of transparent sheets each containing parts of the drawing (Figure 3.2). Whole layers can be manipulated to change the color or linetype for all the objects on a particular layer. Layers can also be made invisible when they are not relevant to the

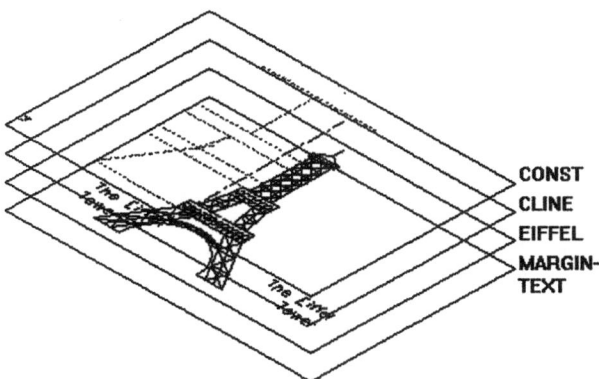

Figure 3.2 Layers as transparent sheets

current task, and then they can be made visible again later. For example, you might use one layer to contain all your construction lines, one layer for text and drawing margins, one for floor plan and another for the wiring diagram. Unlike a conventional paper drawing there is no need to delete your construction lines, simply make them invisible. In this way, when you come to edit the drawing or add in a plumbing diagram, all the original constructions are available.

Because of the importance of layers in the organisation of drawing information and the increasing need for drawing exchange via CAD, most users adopt some convention for using layers. For construction drawings, British Standard 1192 Part 5 sets out recommendations for naming layers and what kind of information should be on them. Because some CAD systems have only a layer numbering facility the standard recommends the names summarised in Table 3.1. For example, the layer S282 would be a layer created by the structural engineer for concrete columns. Readers are referred to the full standard for exact details. This highlights the importance of adopting a logical approach to organisation of information in CAD. Adherence to a common standard allows engineers using different CAD systems to communicate effectively.

When you start a new drawing AutoCAD always has a minimum of one layer called "0" (zero). Extra layers can be set up at any stage during the drawing but it is advisable to decide on the structure of the layers before commencing actual drawing. In older versions of AutoCAD only one linetype and one color per layer were possible. Thus each layer has associated with it a color and linetype. Releases 10 and upwards allow multiple colors and linetypes on a layer but still allow you to give a layer specific default settings.

As an example, our drawing, EXPRESS2, will contain some doodling as you try out some drawing commands, some text, some construction lines and

Table 3.1 Recommended layers in construction drawings

Field 1	Discipline	Field 2	Type of information
A	Architect	100–199	Ground, Substructure – General
B	Building Surveyors	200–299	Structure, Primary Elements
C	Civil Engineers	300–399	Structure, Secondary Elements
D	Drainage Engineers	400–499	Finishes
E	Electrical Engineers	500–599	Services – General
F	Facilities Managers	600–699	Services – Electrical
K	Client	700–799	Furniture and Fittings
S	Structural Engineer	900–999	External Works

Based on British Standard 1192 Pt5 and the Autodesk Ltd & AutoCAD
Users Group Layer Naming Convention for CAD in the Construction Industry

the tower outline. The following sequence gives the AutoCAD prompts and
your responses.

Pick **Settings** from the menu bar followed by **Layer Control. . .** as shown
in Figure 3.3.

This brings up the Layer Control dialogue box, Figure 3.4. Initially, the
layer name list on the left will contain only one value, namely 0. The line
near the top indicates the "current" layer or that to which new entities will
belong. The various buttons and options will be explained below. Only buttons
with black text can be used. The grey text means that those buttons are not
available for the current operation.

To create the new layers shown in Figure 3.4 move the cursor arrow to
the input box just above the OK button. Then type the names of the layers
separated by commas. As the list exceeds the space in the box the letters will
scroll to the left. If you make a mistake in typing use the left and right arrow
keys on the keyboard to move about and type the correction. When all the
names have been typed click the **New** button. They should then be added to
the Layer Name list as shown.

Layer names can contain letters and numbers and have up to 31 characters,
though it is advisable to restrict them to as few as practicable. The names
should be descriptive or follow some conventional numbering or acronym. New
layers can be added at any time.

When AutoCAD creates new layers they have a number of default properties. Their "State" is "On", that is, they are visible. They have a continuous
linetype and the layer color for entities is white. Note that entities will only
be white if your screen is configured to have a dark background. If you have

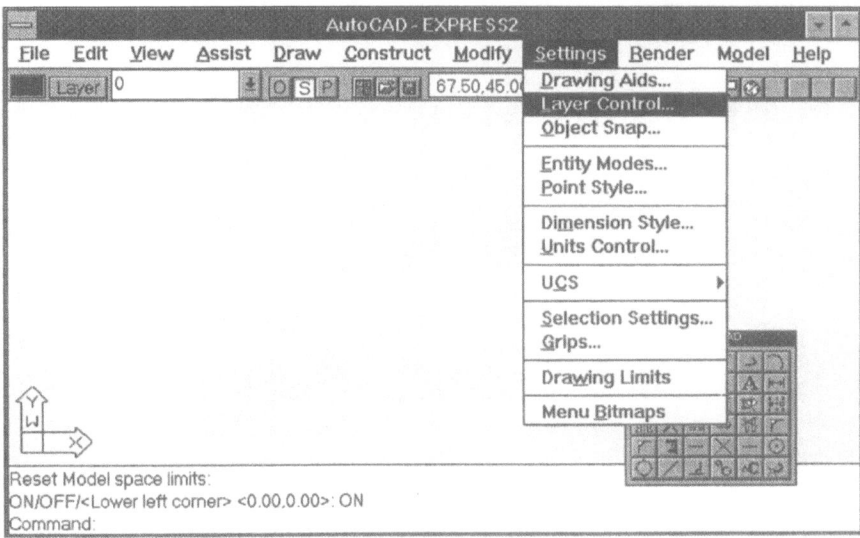

Figure 3.3 Settings/Layer Control... pull-down menu

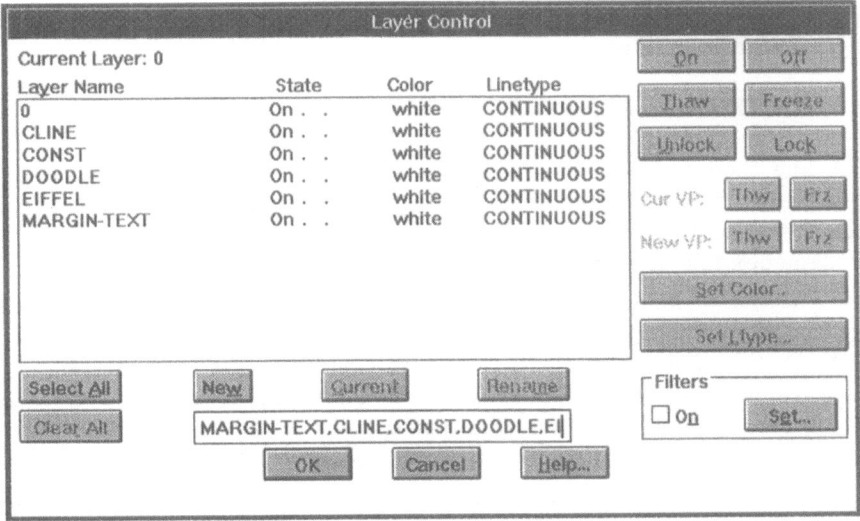

Figure 3.4 Layer Control Dialogue Box

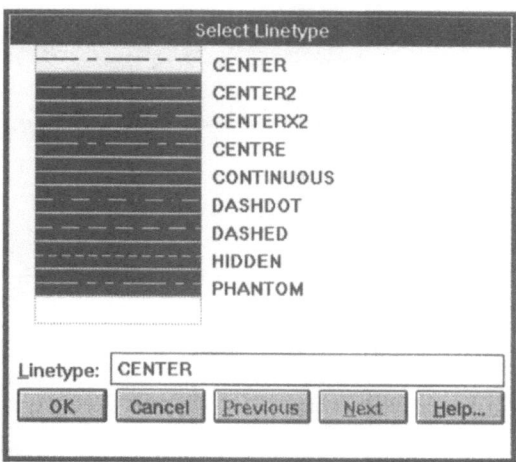

Figure 3.5 Select Linetype dialogue box

a light background then "white" actually means black. White lines normally convert to black lines when plotted on paper.

As we have already loaded the centre linetype we can assign it to the layer CLINE. We can also change a few layer colors. Move the arrow cursor into the Layer Name list and pick the line

CLINE On.. white CONTINUOUS

This line should appear in reverse video. Some of the buttons to the right of the Layer Control dialogue box change from grey to black text indicating that they can now be used. Pick **Set Ltype....** This brings up the Linetype dialogue box (Figure 3.5).

The list of loaded linetypes with their graphic equivalent might differ slightly on your computer. You should at least be able to see CENTER, CEN-TRE, DASHED and HIDDEN. Note that CENTER and CENTRE have the same graphic. Move the cursor arrow to the **CENTER** graphic and click. This should become highlighted and the text appears in the box near the bottom. Pick **OK** to set this and return to Layer Control.

The next operation is to set the color for the construction lines to be red and the doodles to be blue. An advantage of assigning specific colors to layers is that it visually identifies which layer an entity belongs to. Setting colors is similar to setting linetypes. Move the cursor into the Layer Name list and pick the line

CONST On.. white CONTINUOUS

You should now see two lines, CLINE and CONST in reverse video. This indicates that the two layers have been selected for the operation. To deselect CLINE pick the line **CLINE On.. white CONTINUOUS**. Make sure that only CONST is shown in reverse video before picking the **Set Color...** button. The color chart in Figure 3.6 appears.

AutoCAD's standard palette of colors consists of red, yellow, green, cyan (greenish blue), blue, magenta and white. These colors may be referred to by name, as above, or by their numbers (1 to 7 in the order listed above). Further colors with numbers 8 to 255 are also available but are dependent on the type of screen you are using. The number of colors in the full palette shown on your screen will depend on your Windows display driver. Obviously, the colors are only meaningful at this stage if you have a color display. However, even if you have a monochrome screen, colors can be assigned anyway, and it is possible to get multi-color plots. You will see the colours as shades of grey on monochrome screens.

It is recommended to stick to the standard set of colors where possible. As many pen plotters are limited to only 4 or 8 colors the full color palette may not be appropriate. However, many color laser printers will support the full 255 colors so you may wish to choose your favourite shade.

Pick the **red** box on the left of the "Standard Colors" (or your favourite shade of crimson) and then pick **OK**.

Repeat this exercise for the layer DOODLE. Deselect **CONST** and pick the line **DOODLE On.. white CONTINUOUS** from the Layer Name list. Pick **Set Color...** and the **blue** box or your favourite shade of ultramarine. Then pick **OK**. The layer settings should then resemble Figure 3.7.

The final layer operation is to set the layer MARGIN-TEXT as the current layer. This is done by deselecting **DOODLE** and selecting **MARGIN-TEXT** as shown in Figure 3.7. Then pick the **Current** button and then pick **OK** .

Of the other options in the Layer Control dialogue box the most useful are the "Freeze" and "Thaw" buttons. Freezing a layer causes it to become invisible. AutoCAD ignores entities on frozen layers and so drafting can be speeded up by freezing layers when they are not required. Thawing a layer makes it visible again. Frozen layers are stored with the drawing when it is saved and can be thawed at any time. This feature will be used later in the chapter.

Layer settings can also be modified using the Command line. Indeed, as you pick the buttons from the Layer Control dialogue box, you should see the commands appearing on the command line automatically. While the dialogue box gives easy access to the settings, sometimes it is quicker to type the command at the keyboard. The following sequence could have been used to set the color for DOODLE to blue.

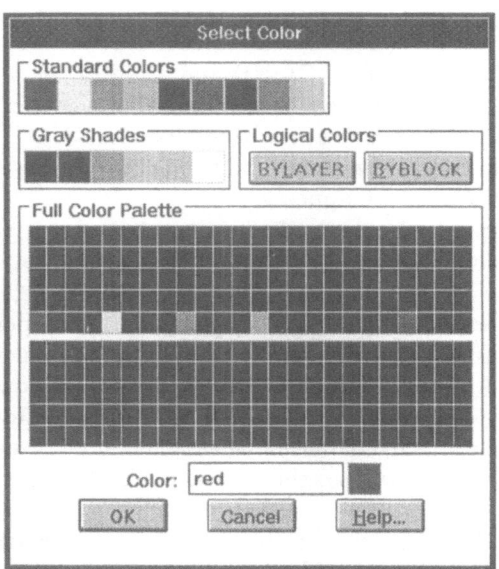

Figure 3.6 Select Color dialogue box

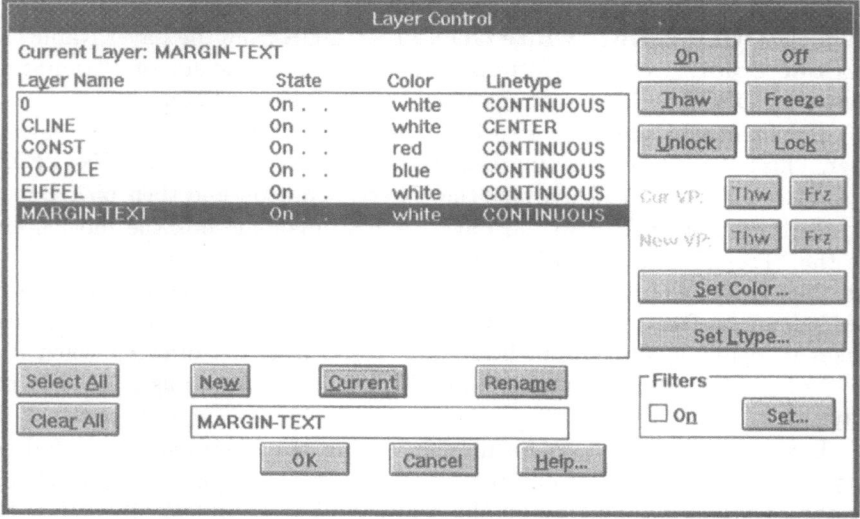

Figure 3.7 Layer settings for EXPRESS1

> Command: **LAYER**
> ?/Make/Set/New/ON/OFF/Color/...: **C**
> Color: **BLUE**
> Layer name(s) for color 5 (blue) <0>: **DOODLE**
> ?/Make/Set/...: <**ENTER**>

This involved five inputs from the user as opposed to six via the dialogue box. When using the command line LAYER returns you to the ?/Make/...prompt. To exit the command, just press enter.

Cursor location

If you move the cursor around the drawing area using the mouse the coordinates on the tool bar line change to give the current location. If the values do not change try pressing the pick button on the mouse. This will give the coordinate of that point. To switch on the continuous read-out of coordinates press ^**D** (press the CTRL key and while doing so press the D key). This should cause the command prompt line to display

> Command:<coords on>

Pressing ^**D** toggle this facility in a circular fashion between dynamic X,Y readout, static X,Y and dynamic polar coordinates. It is recommended that the dynamic X,Y coordinate display is switched on. The **F6** key also does this operates this toggle. The dynamic polar display only operates when rubber banding is in operation eg drawing lines.

Rulers, grids and snapping

In addition to coordinate display AutoCAD gives a grid facility which helps you to get your general bearings within the drawing, and a snap function which helps to pin-point locations exactly. GRID causes an array of equally spaced dots to apppear. These are analogous to placing a sheet of graph paper as a guide beneath the tracing sheet on a drawing board. The grid points are not part of the drawing and do not appear on plots.

The SNAP command is one of AutoCAD's most powerful (and time saving) facilities. By turning SNAP on you can restrict the movements of the cursor to discrete steps. The cursor jumps from one snap point to the next missing out all the points in between. This is particularly useful if you are drawing an item whose smallest dimension is 5 units. You could set SNAP to a value of 5. This makes point picking with the mouse a lot easier and eliminates the need to type the coordinates when exact locations are required.

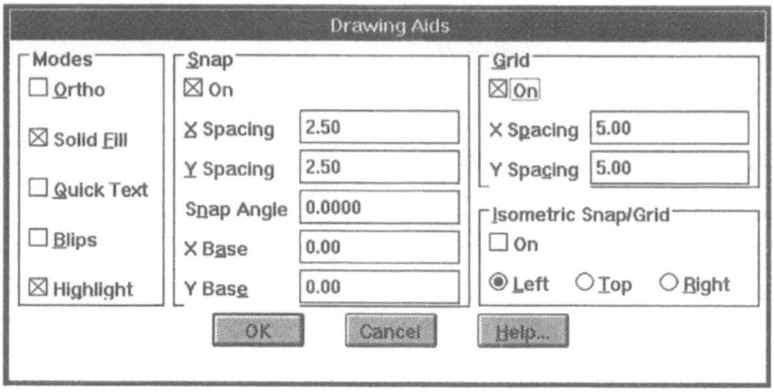

Figure 3.8 Drawing Aids dialogue box

The drawing, EXPRESS2, will be created with a SNAP value of 2.5. The GRID will be set to 5 units. This gives a nice relationship between the two settings. Every second snap point will also be a grid point. If the grid is too dense it can get in the way and slow down the graphics. The best way of setting this up is to pick **Settings** from the menu bar and **Drawing Aids...** from the pull-down menu. This gives the Drawing Aids dialogue box, Figure 3.8.

The middle section of the dialogue box concerns the SNAP settings. Move the cursor to the box beside **X Spacing** click and overtype the current value with **2.5**. If you click on the **Y Spacing** box it will automatically update to be the same as the X value. Make sure that the **On** box is ticked. If it is empty pick the On box with the cursor to activate SNAP. Similarly ste **Grid** to be **On** with X and Y Spacing of **5**. When your box matches Figure 3.8 pick **OK**. The display should then look like Figure 3.9.

When using these features, try to chose sensible spacings. If the spacing is small, the redraw speed will slow down, as AutoCAD has to re-display all the dots and ticks. If the spacing is too small you will get a message "Grid too dense to display". If this happens just re-issue the command and set new spacings.

The tool bar in Figure 3.9 shows the SNAP button. When this is highlighted, it indicates that snapping is currently active. Picking this button will toggle SNAP on and off. Pressing ^B also operates the toggle. There is no toggle button for the GRID but ^G works at the keyboard.

The GRID has a useful side effect in that it shows the limits of the drawing when Zoom All has been activated. It doesn't matter which layer you are on when activating GRID or SNAP. They act as an overlay on the whole drawing. They can be switched on and off as many times as required and can be used

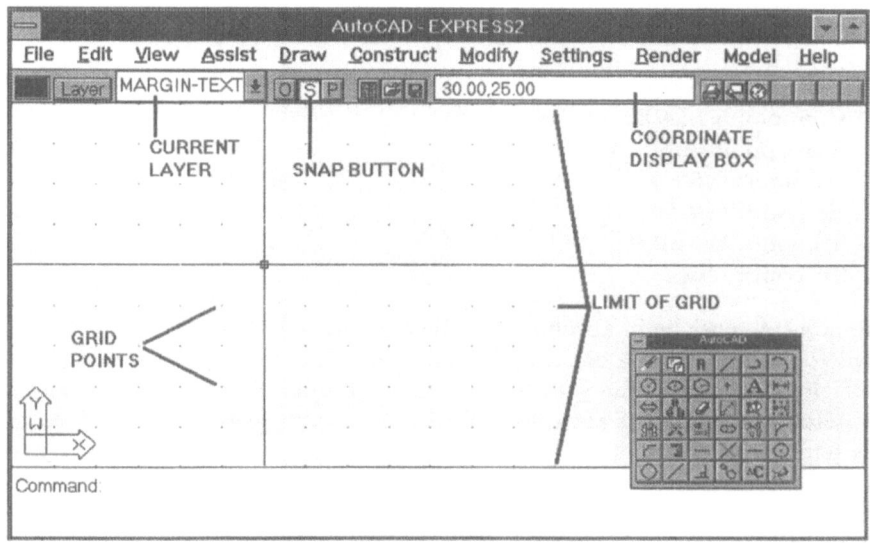

Figure 3.9 Snap and grid

independently of each other. Remember, the **GRID** will not appear on your print out or plot.

Two Dimensional Coordinate notation

You have already used the X,Y coordinate notation. In the last chapter the X was seen to signify horizontal distance from the origin and Y vertical distance. A point on the drawing can be located by keying in a pair of numbers separated by a comma. For example, the coordinates "4,3" belong to a point 4 units to the right and 3 units above the origin. The origin has the coordinates "0,0". Mathematics books call this the Cartesian coordinate system after the French philosopher and mathematician, Descartes. AutoCAD calls it the X-Y WORLD system (WORLD meaning that the values are in relation to the drawing origin).

Sometimes it is more convenient to work in distances relative to one's current location. For example, when giving directions to a stranger in town your instructions might be "Follow this road for half a kilometre and turn right. The Computer Training Firm's offices are a further 350m on the left." This is a lot more meaningful than giving the location in terms of a map grid reference. Similarly, you can tell AutoCAD to draw a line giving directions relative to the cursor location. The notation used in AutoCAD to signify *relative* coordinates is to put "@" in front of the X,Y pair of values. For example to draw margins for EXPRESS2 try the following:

Command: **LINE**
From point: **2.5,2.5**   (A)
To point: **@60,0**   (B)
To point: **@0,40**   (C)
To point: **@−60,0**   (D)
To point: **close**

The point A must be in absolute X-Y WORLD coordinates since it is the first point to be input to the drawing. The second point, B, is given as 60 units to the right of A and at the same height. C is 40 units directly above B and so on. Relative coordinates always use the immediately preceding cursor location as a temporary origin.

If you know the length of a line and the angle it makes with the horizontal then it can be easier to use *relative polar* coordinates. To draw a vertical centre-line from point E (30,5) which is 36.5 units long first switch layers and you use the LINE command again.

The easiest way to switch layers is to pick the **down pointing arrow** on the right of the current layer box on the tool bar, Figure 3.10. This gives a pull-down menu listing all the available layers. Pick **CLINE** and the change is executed. Both the Current Layer and Color Box should change to CLINE and red. Now draw the line.

Command: **LINE**
From point: **30,5**   (E)
To point: **@36.5<90**   (F)
To point: **<ENTER>**

The "<" or "less than" character indicates that the next character is an angle. Thus the above line will be drawn at 90 degrees from the horizontal. As positive angles are anti-clockwise the line goes up and not down. Again the "@" means that the location is relative to the previous point (30,5). You can use *absolute polar* coordinates when the angles relate to lines from the origin to the new location. This can be cumbersome and is not generally recommended.

If you want to experiment with relative and absolute cooordinate notation, try drawing some objects on the DOODLE layer. On this doodle detour of the AutoCAD Express you can take in some sights on the scenic route. If you wish to press ahead without further practice you can skip forward to the next section entitled "Digital setsquares".

The pyramids of Egypt can be drawn as triangles. First, switch to the DOODLE layer. This can be done using the tool bar or by typing LAYER at the Command: prompt

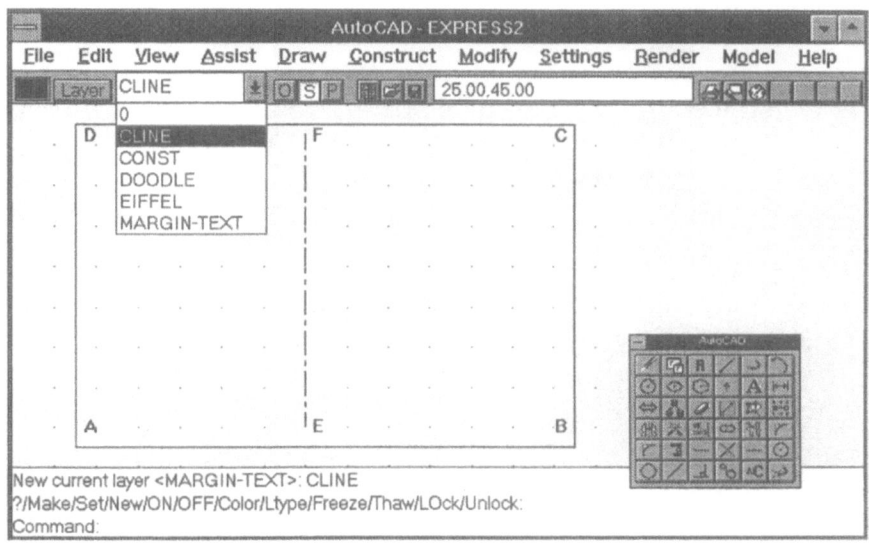

Figure 3.10 Margins and centre-lin

Command: **LAYER**
?/Make/Set/...: **S**
New current layer <0>: **DOODLE**
?/Make/Set/...: **<ENTER>**
Command: **LINE**
From point: **10,30**   (P1)
To point: **@5,6**   (P2)
To point: **@5,−6**   (P3)
To point: **@10<180**   (P1)
To point: **<ENTER>**
Command: **LINE**
From point: **5,15**   (P4)
To point: **@12<60**   (P5)
To point: **@12<−60**   (P6)
To point: **close** or Pick Assist pull-down menu and Close

Experiment with other shapes and combinations of the different coordinate notation.

Now is as good a time as any for saving the drawing. Remember the motto "Save early and often".

Command: **SAVE**

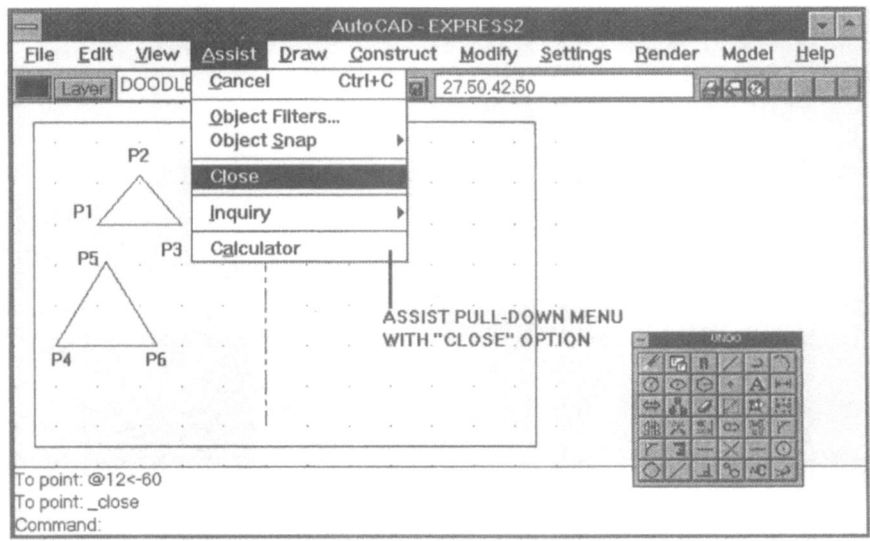

Figure 3.11 The pyramids of Egypt

File name <EXPRESS2>: <ENTER> or pick **OK** from the dialogue box.

Digital setsquares

A large part of any drawing consists of horizontal and vertical lines such as the margins and centre-lines above. AutoCAD's ORTHO command allows you to draw these horizontal and vertical lines quickly and with 100% accuracy. When the ORTHO mode is ON then all movements of the cursor are restricted to either the X direction or the Y direction.

The simplest way to switch ORTHO on is to click the **O** button on the tool bar (Figure 3.11). You can also toggle ortho by pressing ^**O** (CTRL and the letter O pressed together) or the **F8** function key on the keyboard. You can also turn it on by typing

Command: **ORTHO** ON/OFF: **ON**

Picking **O** is easier and pressing it a second time switches back to normal drawing mode. When it is active the word "ORTHO" appears on the status line at the top of the screen and as you switch it on or off, the command prompt area will echo your action with "<ORTHO ON>". To examine these effects draw the horizontal construction lines outlined below for EXPRESS2 making sure that ORTHO is ON.

Use the command line or **Settings/Layer Control...** to set the **Current** layer to CONST and to **Freeze** the DOODLE layer .

> Command: **LAYER**
> ?/Make/Set/...: **S**
> New current layer <0>: **CONST**

Freeze any doodles that might get in the way. Note DOODLE could not be frozen while it was the current layer.

> ?/Make/Set/...: **F**
> Layer name(s) to freeze: **DOODLE**
> ?/Make/Set/...: **<ENTER>**
> Command: **LINE**
> From point: **17.5,5**   (G)

Now move the cursor to the right of the centre-line (EF) near to G'. Even though the cursor is not exactly horizontally across from the point G, the line from it is (Figure 3.12). If you move the cursor back towards G there is a point when the line suddenly jumps to being vertical. So, wherever you pick the point the line is restricted to the X and Y directions. The governing factor for whether it is vertical or horizontal is the larger magnitude, X or Y, from the first point to the cursor.

Move the cursor back to G' and pick the end point.

> To point: pick a point to make a horizontal line   (G')
> To point: **<ENTER>**
> Command: **<ENTER>**
> **LINE** From point: **17.5,15**   (H)
> To point: pick a point to make a horizontal line   (H')
> To point: **<ENTER>**

Do the same for horizontal lines from points J (17.5,17.5), K (17.5,22.5), L (17.5,25) and M (17.5,38). Your drawing should now resemble Figure 3.13 (without the letters). Your construction lines should be solid and red. They are shown as dashed in later figures for clarity.

Now is a good time to save your drawing. You can save it as many times as you like, during a session. You will get a message stating that the file already exists. Pressing **<ENTER>** or **OK** will overwrite the previously saved version.

> Command: **SAVE** File name <EXPRESS2>: **<ENTER>**

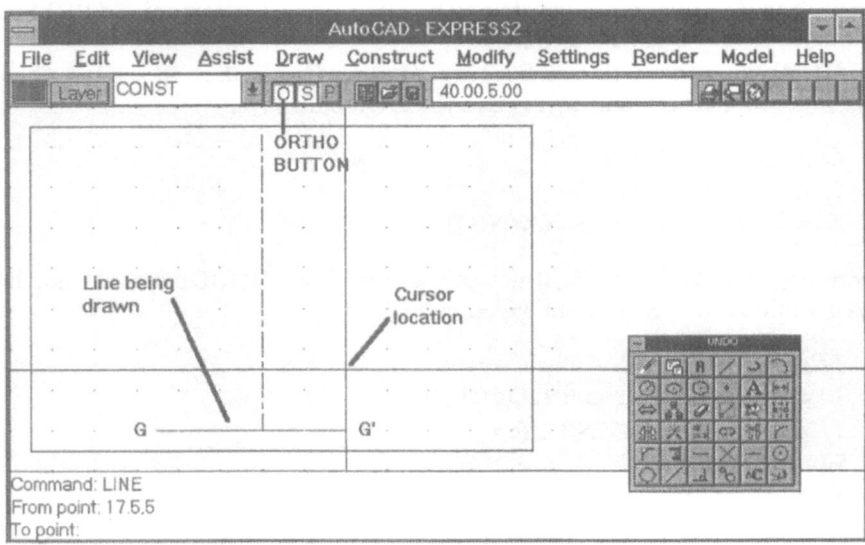

Figure 3.12 ORTHO mode

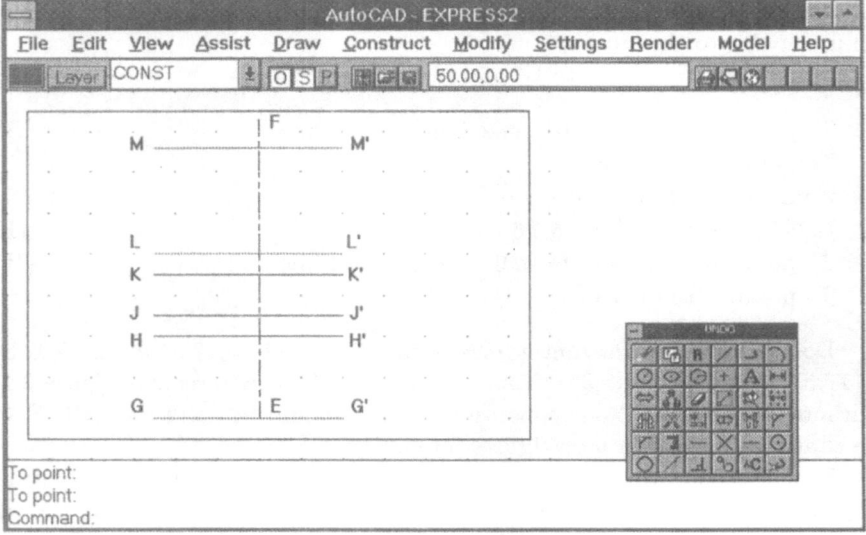

Figure 3.13 Horizontal constuction lines

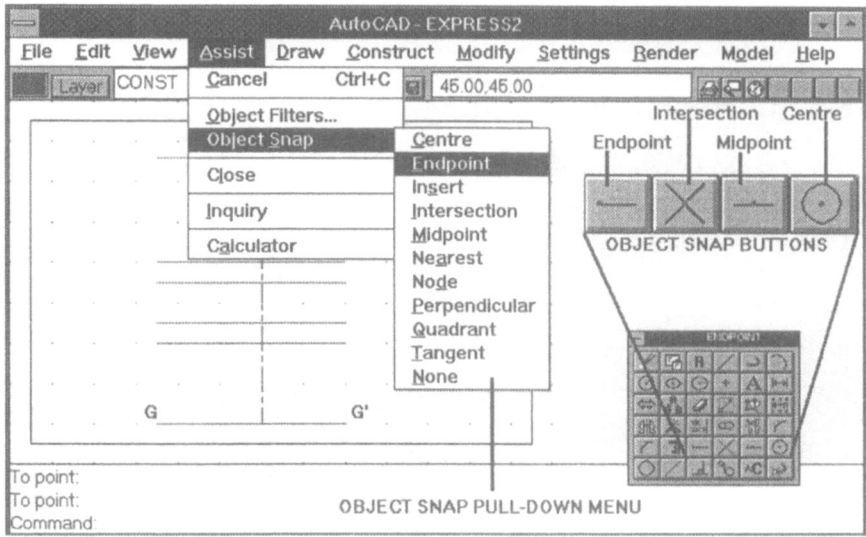

Figure 3.14 Object Snap menu and buttons

Snapping to objects

The main usefulness of construction lines is that by drawing them you reduce the number of calculations necessary to locate awkward points. The intersection point between a line and a circle is easy to draw but may require clever geometry to calculate. This is not to suggest that the AutoCAD user is not clever. Rather, you just want the fastest and simplest solution!

AutoCAD allows you to snap to key points of previously drawn items. The Object SNAP pull-down menu can be found by picking **Assist** from the menu bar. This menu gives you facilities to locate the centre or tangent point of a circle or arc, the end point and mid point of a line, the insertion point of text etc. The more commonly used object snap functions are also included in the bottom two lines on the floating tool box. Some of these are indicated in Figure 3.14.

There are two ways of using Object Snap. Firstly, it can be employed to find a single point by invoking it in response to AutoCAD prompts requesting a point (eg "From point:"). The next few lines illustrate this feature. Type **LINE** and at the "From point:" prompt pick the **Endpoint** button in the tool box (third from left on second last row). The prompt changes to "end of:" and the cursor changes shape to give a larger target box. Move the cursor

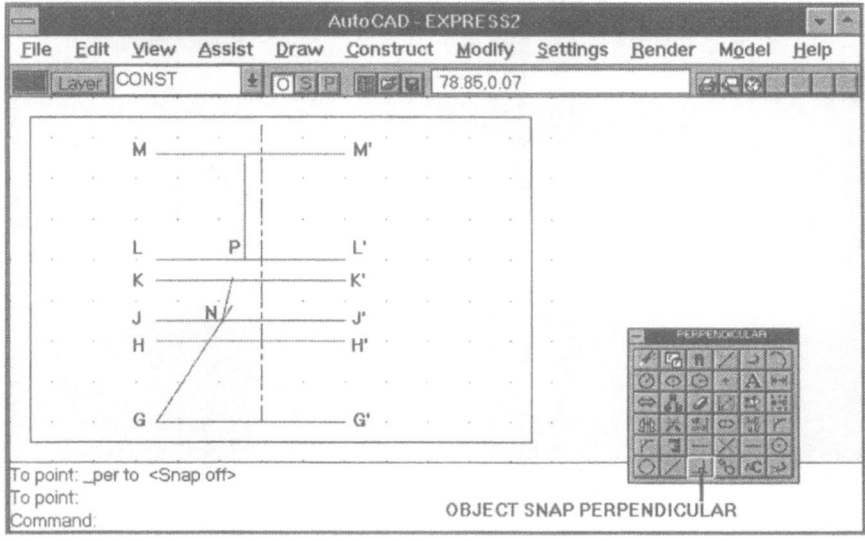

Figure 3.15 Using OSNAP options

until you are near point G on the line GG'. Once it is in your sights, pick the point.

> Command: **LINE**
> From point: _end of: pick line near G
> To point: **@17<58**
> To point: **<ENTER>**

To draw a line from the intersection point of the last line and the line JJ' use **Intersec** in reply to "From point:" and move the cursor so that point N is within the target box.

> Command: **<ENTER>**
> LINE From point: Int of: pick point N
> To point: **@5.5<77**
> To point: **<SPACE BAR>**

Remember, the space bar acts just like the enter key. Now to draw the vertical line on Figure 3.15 from the point, P, and perpendicular to the line MM', use tool box "perpendicular" button on bottom line. Use "^B" or pick "O" from the tool bar to toggle the snap mode.

> Command: **<SPACE BAR>**

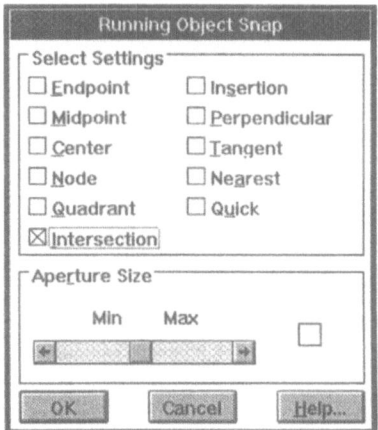

Figure 3.16 Running Object Snap dialogue box

LINE From point: **28,25**   (P)
To point: **PERP** to ^**B** <Snap off> Pick line MM'
To point: ^**B** <Snap on> <**ENTER**>

This way of using Object Snap, or OSNAP as the AutoCAD command is called, gives a single point selection and then returns to normal selection straight away. It can be used only when AutoCAD is expecting you to pick a point.

When you want to connect up a lot of construction points using Object Snap it can become tedious to have to pick, say, Intersection each time. The second way of using Object Snap is to set up a continuous "OSNAP" mode. This is done by picking **Settings/Object Snap...** from the menu bar. This birngs up the Running Object Snap dialogue box, Figure 3.16. Pick **Intersection** and then **OK**.

This means that AutoCAD will always snap to the intersection point nearest to the centre of the target box. Note that Figure 3.16 also has a section for controlling the "Aperture Size" or the size of the target box. The slider bar can be used to either increase or decrease the size. The smaller the aperture the faster AutoCAD will locate the point. However, the smaller size makes it more difficult for the user to pick entities.

Now, use this to draw the outline of the tower (Figure 3.17). First change layers. Do this by picking the **down arrow** next to the layer name box on the tool bar and then pick **EIFFEL** .

Command: **LINE:**
From point: **22.5,5**   (Q)

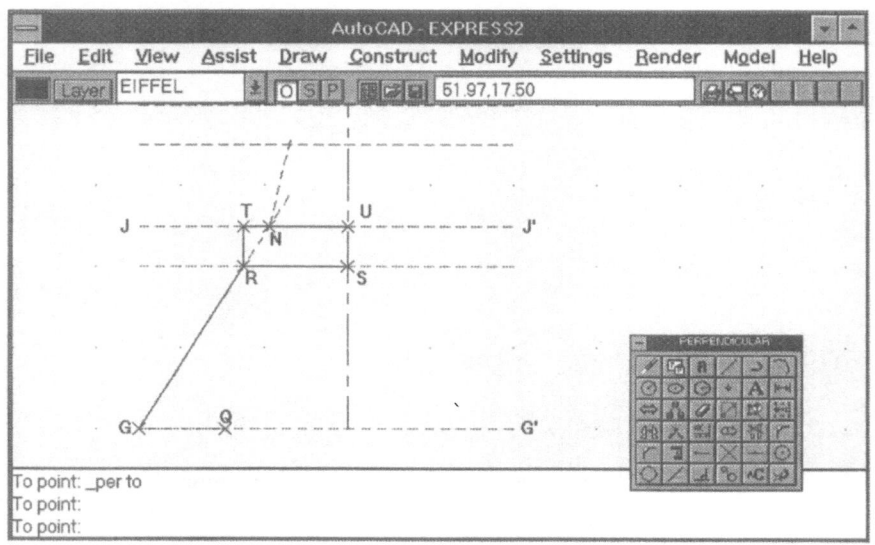

Figure 3.17 OSNAP mode INTersec

> To point: **@−5,0** or pick the point, G   (G)
> To point: Pick R (Object snap overrides ORTHO)
> To point: **<ENTER>**
> Command: **<ENTER>**
> LINE From point: Pick S
> To point: Pick R

Now to draw a line from R perpendicular to the line JJ' select **PERPend** from the OSNAP menu. This overrides INTersec for the next point selection.

> To point: perp to: Pick line JJ'   (T)
> To point: Pick U
> To point: **<ENTER>**

This procedure is repeated to draw the remainder of the left hand outline (Figure 3.18).

> Command: **LINE:**
> From point: Pick N
> To point: Pick V
> To point: **<ENTER>**
> Command: **<ENTER>**
> LINE From point: Pick W

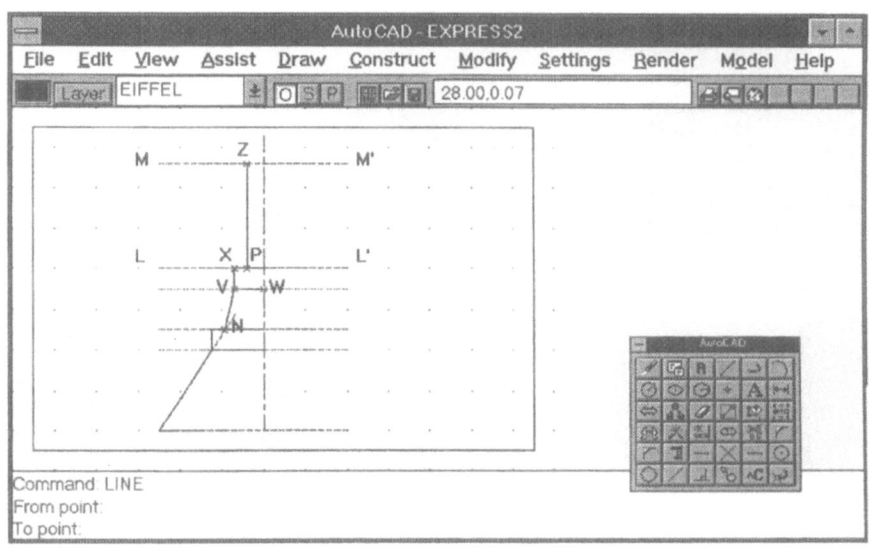

Figure 3.18 Half an Eiffel

To point: Pick V
To point: **perp** to: Pick line LL'   (X)
To point: Pick Y
To point: <**ENTER**>
Command: <**ENTER**>
LINE From point: **28,25**   (P)
To point: Pick intersection point on MM'   (Z)
To point: <**ENTER**>

Now to turn the OSNAP mode off. Pick **Settings/Object Snap...** and then click the **Intersection** box to remove the X and pick **OK**. An alternative way of doing this is to use the OSNAP command as follows.

Command: **OSNAP**
Object snap modes: **NONE**

As you pick points on the screen, little blips or crosses may appear. The blip gives you some visual feedback about the location of the point and it is easy to see if it is at the correct location. However, the more points you pick the greater the distraction caused by these blips. To clear the screen of old blips you must execute the **REDRAW** command. This simply redraws the current display and as the blips are not true parts of the drawing they do not reappear.

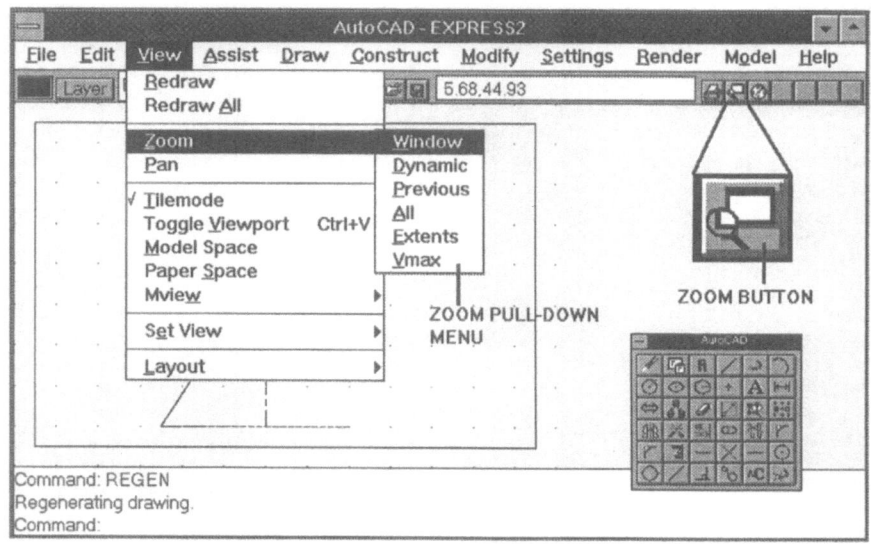

Figure 3.19 Zoom pull-down menu and button

REDRAW is in the VIEW pull-down menu and also appears as the letter "R" in the top row of the tool box.

Command: **REDRAW**

ZOOM and PAN

You have already used the "ZOOM All" command to display the whole drawing area. In this section you will use ZOOM to enlarge the view of the top of the tower being drawn so that greater detail can be added to it. The best way to imagine how zooming works is to think of the computer display as the image you would see looking through a camera lens. As you adjust your zoom lens the item you are looking at becomes enlarged while peripheral items are excluded from view. The PAN command is another term from the camera man's vocabulary. It allows you to sweep your "camera lens" over the drawing to look at other parts with the same magnification.

To enlarge the top of the tower pick **View** from the menu bar and then **ZOOM**. Then pick **Window** from the ZOOM menu, Figure 3.19. You are asked to give the two opposite corners of the new window of vision. There is also a ZOOM button on the tool bar.

Pick the points W1 and W2 shown on Figure 3.20.

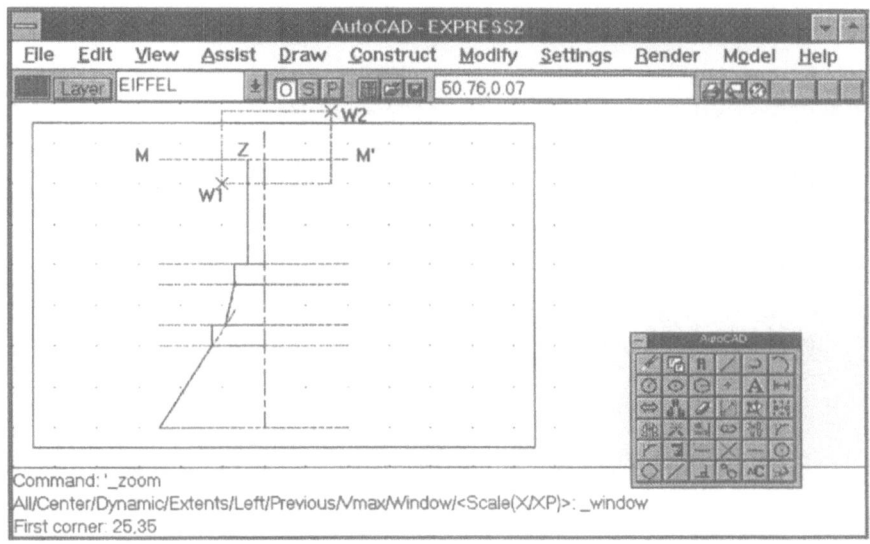

Figure 3.20 Picking a window

Command: '_zoom
All/Centre/.../<Scale(X/XP)>: _window
First corner: **25,35**   (W1)
Other corner: **38,44**   (W2)

If you pick the ZOOM button you will have to type the "W" or "window" at the "All/Center..." prompt and then pick the two points. If you pick View/Zoom/Window from the pull-down menu then you just need to pick the two points.

The display should now change to give a close-up view of that rectangular window (Figure 3.21). The spacing between the grid points and axis ticks looks larger but the coordinates are just the same. As you move the cursor around the drawing the read-out on the status line will give the coordinates with respect to the drawing origin. Nothing has changed in the drawing, it's just the amount you can see that has altered.

If your display is significantly different from Figure 3.21 then the wrong window must have been chosen. Do a ZOOM All and repeat the ZOOM Window, typing in the coordinates for W1 and W2.

As you are working at a larger scale to draw small items you can change the GRID and SNAP settings. This can be done with **Settings/Drawing Aids...** or by typing the commands. If you type the commands there is an extra facility available.

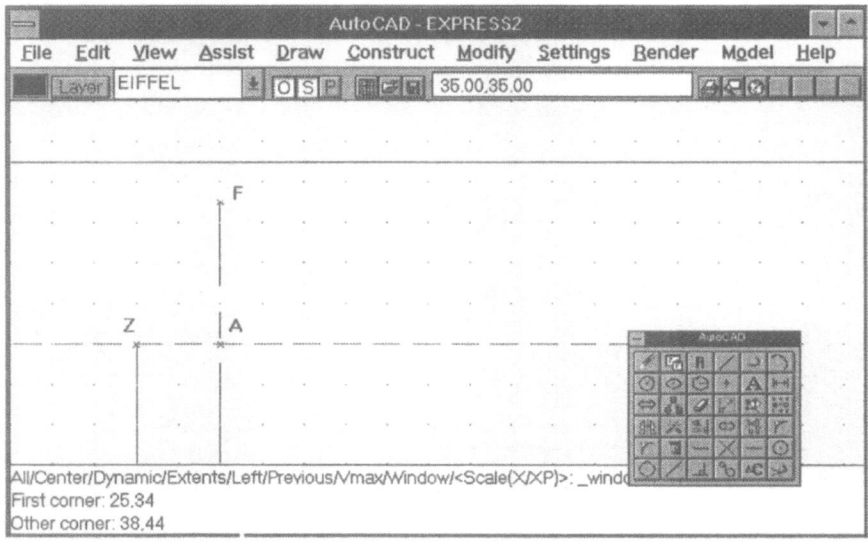

Figure 3.21 The enlarged image

> Command: **SNAP**
> Snap spacing or ON/OFF/Aspect/Rotate/Style <2.50>: **0.5**
> Command: **GRID**
> Grid spacing or ON/OFF/Aspect/Snap/Style <5.00>: **2X**

The "2X" sets the grid to twice the current snap setting.

Now draw the viewing platform with its roof and aerial (Figure 3.22). Use the **Intersection** button on the tool box for points A and Z.

> Command: **LINE**
> From point: **INT** of: Pick A
> To point: **@−2.5,0**   (B)
> To point: **@0,−1**   (C)
> To point: **@2.5,0**   (D)
> To point: **<ENTER>**
> Command: **<ENTER>**
> LINE From point: **INT** of: Pick Z

You can switch ORTHO off in the middle of the LINE command by picking **O** from the tool bar and then pick the point E (30,39.5), or you can type the coordinates.

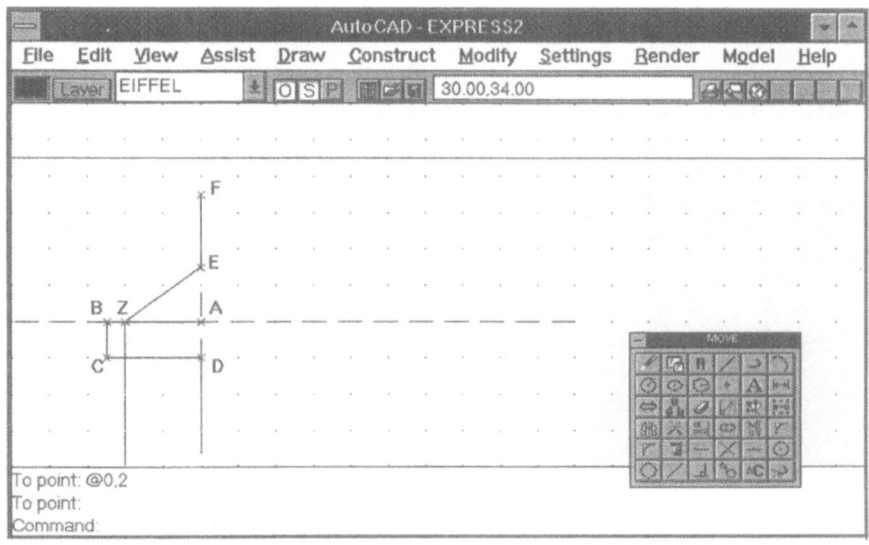

Figure 3.22 Half the viewing platform

To point: <ORTHO OFF> **30,39.5** (E)
To point: **@0,2** (F)
To point: **<ENTER>**

Now ZOOM in even closer using window to add some structural detail.
Pick the **ZOOM button** with the magnifying glass from the tool bar.

Command: ZOOM
All/Centre/.../<Scale(X/XP)>:**W**
First corner: **27,35.5**
Other corner: **30.5,39**

The magnification is further increased and using the same SNAP resolution
you can draw the vertical line GH and the horizontal line JK, as shown in
Figure 3.23.

Command: **LINE**
From point: **29,38** (G)
To point: **@0,−1** (H)
To point: **<ENTER>**
Command: **<ENTER>**
LINE From point: **27.5,37.5** (J)

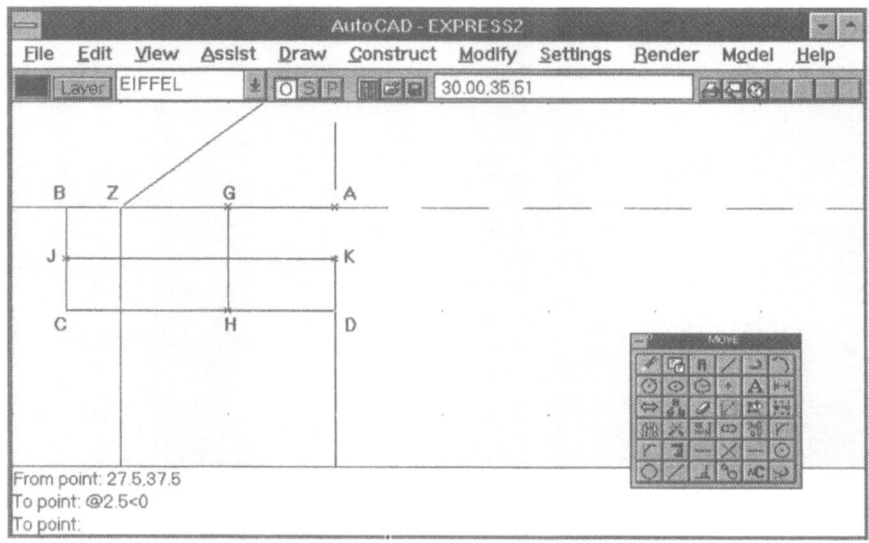

Figure 3.23 Adding some structure

To point: **@2.5<0**   (K)
To point: **<ENTER>**

To ZOOM back to the last magnification pick **View/Zoom** and then pick **previous** or type:

Command: **ZOOM**
All/Centre/.../<Scale(X/XP)>: **P**

AutoCAD stores the settings of up to ten previous zoomed views and so you can step backwards that many times. This facility saves you having to do a **ZOOM All** followed by a new ZOOM Window to get back to an earlier view. Your picture should now resemble Figure 3.22 but with the new lines in place.

Another feature which saves the double task of zooming out and back in is the PAN command. To see something that is currently out of view you can PAN the imaginary camera. PAN is also available on the View pull-down menu. For example, to see the middle landing of the Eiffel tower under construction try the following sequence.

Command: **PAN** Displacement: **2,15**
Second point: **<ENTER>**

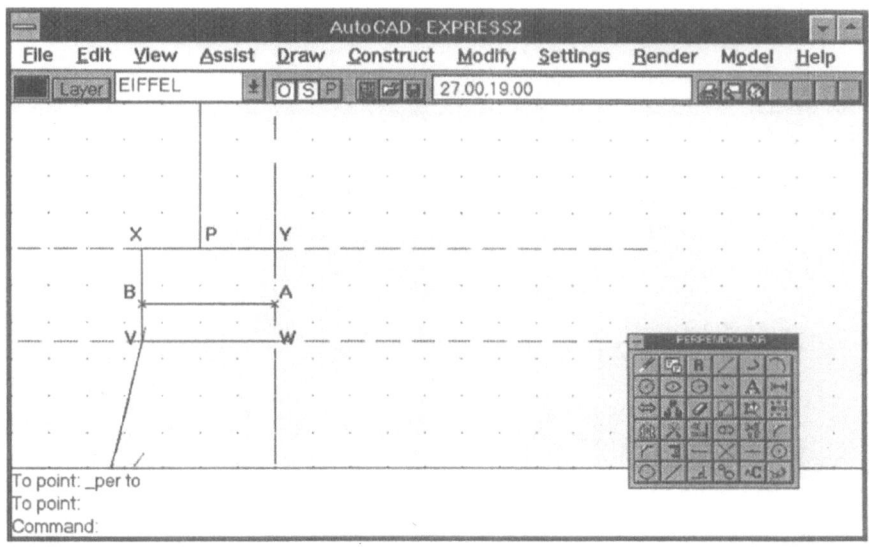

Figure 3.24 Middle landing after PAN

This moves the drawing two units to the right and 15 units up. If you simply press <ENTER> in reply to the "Second point:" prompt then the first pair of coordinates are taken as a relative displacement. The same movement could have been achieved by entering

Command: PAN Displacement: 30,37
Second point: 32,52

The second method gives a vector which controls the movement. The coordinates can also be input by picking points on the screen. However, as the point 33,52 does not appear on the screen it has to be typed. Remember, it is only the display that is "moving"; the drawing retains all its original coordinate information.

This middle landing of the tower needs one more line. Draw the line AB using ORTHO and Object Snap perpendicular (Figure 3.24).

Command: **LINE**
From point: **30,23.5**   (A)
To point: **PERP** perpend to pick line VX   (B)
To point: <ENTER>

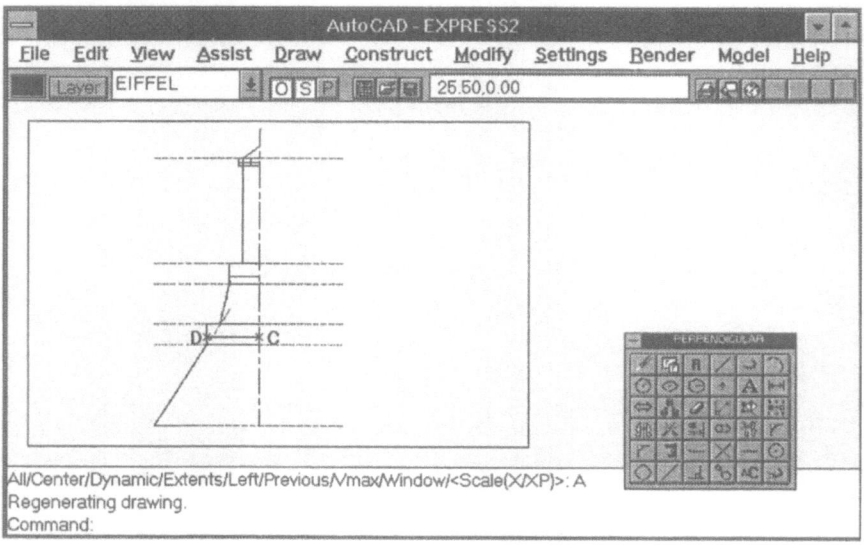

Figure 3.25 Back to full view

To see the lower landing pick **PAN** once more and then pick the points
P1 (30,20) and P2 (32,25) and draw the line CD given below. You may need
to toggle SNAP to off with ^B in order to pick the line at D.

Command: **PAN** Displacement: **30,20**   (P1)
Second point: **33,27**   (P2)
Command: **LINE**
From point: **30,16**   (C)
To point: **PERP** perpend to pick vertical line at D (Figure 3.25)   (D)
To point: **<ENTER>**

Finally,

Command: **ZOOM**
All/Centre/.../<Scale(X/XP)>: **A**

and the display should match Figure 3.25.

The picture may not look like much at the moment, but if you skip ahead
a few pages you can see its potential. You will need to use **EXPRESS2** again
in Chapter 5, so make a safe backup copy.

Command: **SAVE**
Pick **"TYPE IT "** in the dialogue box

File name <EXPRESS2>: **TOWER**

Make a copy onto a floppy disk as well. Place a formatted diskette in Drive A: and do another **SAVE**.

Command: **SAVE**
Pick "**TYPE IT** " in the dialogue box
File name <EXPRESS2>: **A:TOWER**

HAZARD WARNING! If you are using the floppy disk drive to store your original drawing files do not remove the floppy disk at any time during the editing session. The default AutoCAD configuration stores temporary files in the same directory as the original drawing. If you remove the disk before exiting AutoCAD then these temporary files will not be erased and may corrupt the disk. When you exit AutoCAD from File the pull-down menu, all temporary files are deleted. Therefore, if you are already using the A: drive then it is safest to make your backup copies after ending the AutoCAD editing session. Use the Windows File manager or Dos.

Keyboard toggles and transparent commands

Before ending AutoCAD for this chapter, here are some descriptions of the various switches or toggles that can speed up your drafting. By pressing the CTRL key in combination with other keys, various display control commands can be executed, even in the middle of doing another command. Some of the toggles cause alterations to the current command. Some of the grey keyboard function keys duplicate these, though there can be differences between makes of computer. Table 3.2 gives a list of the main keyboard CTRL combinations.

A transparent command is one that can be run while another is still being executed. For example, the ORTHO, ^O, command was issued above in the middle of a LINE sequence. This type of transparency of the ORTHO command greatly increases the flexibility with which it can be used. Similarly SNAP and GRID can be used transparently by pressing the CTRL code or function key. The single point usage of OSNAP is another example.

The SNAP, GRID, and ORTHO settings can also be altered transparently from the pull-down menus and dialogue boxes or switched on/off using the toggles.

REDRAW, ZOOM and PAN can run during other commands if they are picked from the View pull-down menu or if they are prefixed by a ' (eg Command: 'ZOOM etc.). The 'ZOOM and 'PAN will only work if the change in magnification is less than a factor of about 10. If it is more than 10 then Au-

Table 3.2 Keyboard Toggles in Windows

Key	Action
^B or F9	Toggles SNAP on and off
^C	Cancels the current command
^D or F6	Dynamic/static cursor location read-out on tool bar
^G or F7	Toggles the GRID on and off
^O or F8	Toggles ORTHO on and off
F1	Executes AutoCAD's Help facility
F2	Toggles text and graphics windows
ESC	Causes pull down menus to disappear

toCAD will probably have to REGENerate the drawing to maintain sufficient screen resolution.

Another transparent command that is useful to know about is the context sensitive help facility. If you are in the middle of a command such as LINE and you want to find out what to do next type **'Help** at any time.

Command: **LINE**
From point: **'HELP**

To exit AutoCAD for Windows Help pick **File** and **Exit**. When you exit the help facility you can continue with the command or cancel it.

Resuming LINE command
From point: **^C**

The AutoCAD HELP information is intended as a reference only and not as a learning aid. Its explanations of commands are thorough and so can appear complicated. Use this book for learning new commands and use HELP as a reminder. In this way you will understand the workings of the commands and be relieved of the overhead of having to remember it all. The online help, 'HELP, is also available through the pull-down menu and appears as a button on dialogue boxes.

One more command that can be executed transparently is LAYER. However, not all of the options within LAYER can be used in this way. Pick **Settings/Layer Control** pull-down menu from the menu bar. Some options such as thawing a layer or changing its linetypes will only be executed when the drawing is next REGENerated and so will not come into force immediately.

Finish for the time being

The final task in this chapter is to END the drawing of the EIFFEL tower and exit AutoCAD for a well deserved break. Pick **File** and **Exit AutoCAD**. Pick **Save Changes** from the Drawing Modification dialogue box if it appears.

Even though you have exited AutoCAD, that is not the end of the EXPRESS2 drawing. The "Exit AutoCAD" just means to end this editing session. You will return to edit it and add in some fancy iron work in Chapter 5. So don't erase it from your disk! And keep a safe backup copy on another disk. If anyone else is learning AutoCAD from this book, make sure you don't mix up each other's drawings.

Summary

In this chapter you have become an expert LINE drawer. In doing so you have encountered all of AutoCAD's drawing aid facilities in one shape or form. When doing an AutoCAD drawing, you need to think in terms of these facilities so that your plan of action makes best use of them and speeds up your drafting. Always start drawings with the key construction lines.

You should now be able to:

Input relative and absolute coordinate locations.
Use polar coordinates for points.
Set up a suitable grid overlay.
Restrict cursor movements to discrete SNAP points.
Restrict cursor movements to ORTHOgonal directions.
Locate the intersection and end points of lines.
Draw lines perpendicular to other line.
ZOOM in to magnify details and ZOOM out to see the whole picture.
PAN across the magnified picture.
Make back up copies of your drawing on a floppy disk.
Run commands transparently.

Chapter 4 DRAWING AND EDITING

General

The AutoCAD Express takes to the air for this chapter's exercise. The bulk of AutoCAD's drawing entities are introduced along with some simple editing. This express air-service will be by hot air balloon (Figure 4.1)!

As you will be using much the same drawing environment as EXPRESS1.DWG it can be substituted as the default drawing. AutoCAD normally uses the drawing ACAD.DWG as the default for all new drawings. That is all the settings and any entities on ACAD.DWG are copied into the new drawing. Before commencing the balloon drawing make sure that you have a copy of your previous drawing EXPRESS1.DWG.

If you don't have a copy of EXPRESS1.DWG handy you could make one up. It should have limits of (0,0) and (65,45) and set an LTSCALE value of 0.1. Now draw some lines. It doesn't matter where or how many but you will have to improvise when erasing them later. Add your name using DTEXT at the point (40,10) with a height of 2 units, zero rotation and then SAVE the drawing.

To use an alternative default drawing start up AutoCAD for Windows. Then pick **File/New** from the menu bar. When the Create New Drawing dialogue box appears enter the new drawing name as **BALLOON** and the prototype name as **EXPRESS1** as shown in Figure 4.2.

If AutoCAD cannot find the prototype drawing file you can search the disk by picking the **Prototype** button. This brings up a file selection dialogue box as shown in Figure 4.3. From this dialogue box you can search other directories and drives. If you still cannot see the drawing file pick the **Find file** button which brings up a further dialogue box. Give the file name, EXPRESS1.DWG and pick **SEARCH**.

Setting the prototype drawing to EXPRESS1.DWG tells AutoCAD to use that drawing as the default and to copy it into the new drawing. The screen should now appear like Figure 4.4 which should be the same as Figure 2.15.

Figure 4.1 Target drawing for chapter 4

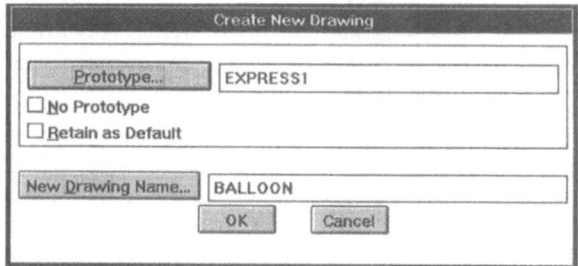

Figure 4.2 Setting the prototype drawing name

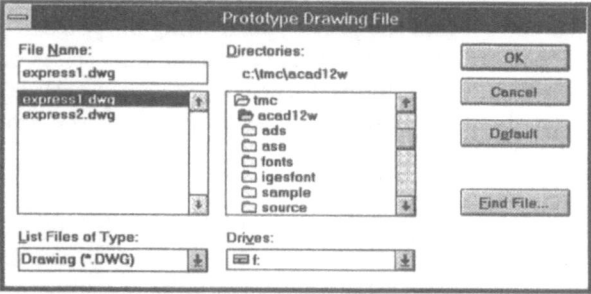

Figure 4.3 Prototype drawing file selection

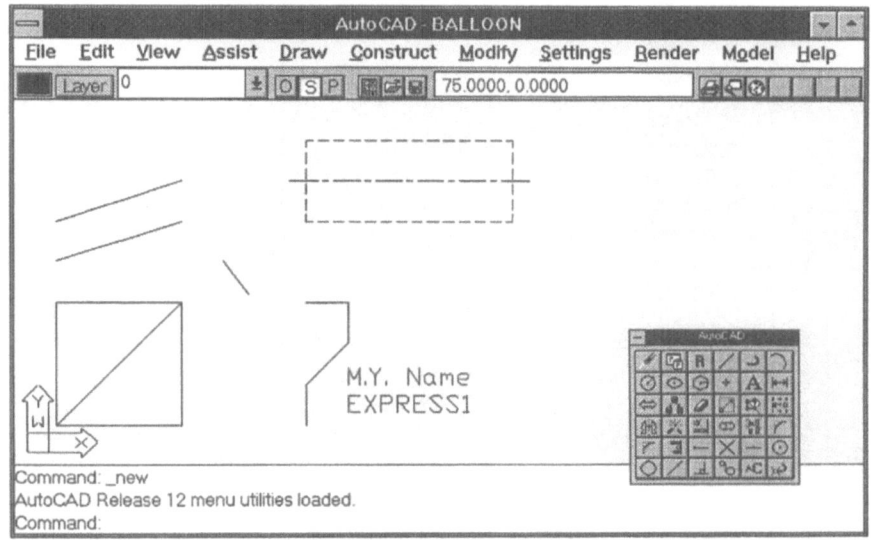

Figure 4.4 Balloon =Express1

Rubbing out and OOPS

When something has been drawn incorrectly or is no longer required on the drawing you can rub it out. This task is a lot easier with AutoCAD than on a conventional drawing board. With AutoCAD's ERASE command you will always get perfect results. The first part of the Balloon exercise is to selectively delete all the objects in the drawing. This will be done in stages to introduce the "Entity Selection" procedure. Once this is accomplished, you can begin to draw the picture in Figure ??.

Entity selection modes

Entities, lines, text, etc. can be selected for editing in one of two modes. Users of earlier versions of AutoCAD will be familiar with the "verb/noun" mode. The command or "verb" is picked first and then the objects, or nouns to be edited are selected. For example, to erase a line one would first pick the ERASE command and then pick the line. With Release 12 and AutoCAD for Windows you can use the more efficient "noun/verb" mode. You pick all the objects first and then pick the command.

To make sure that noun/verb selection has been enabled pick **Settings** and **Selection Settings....** In the dialogue box, Figure ??, make sure that

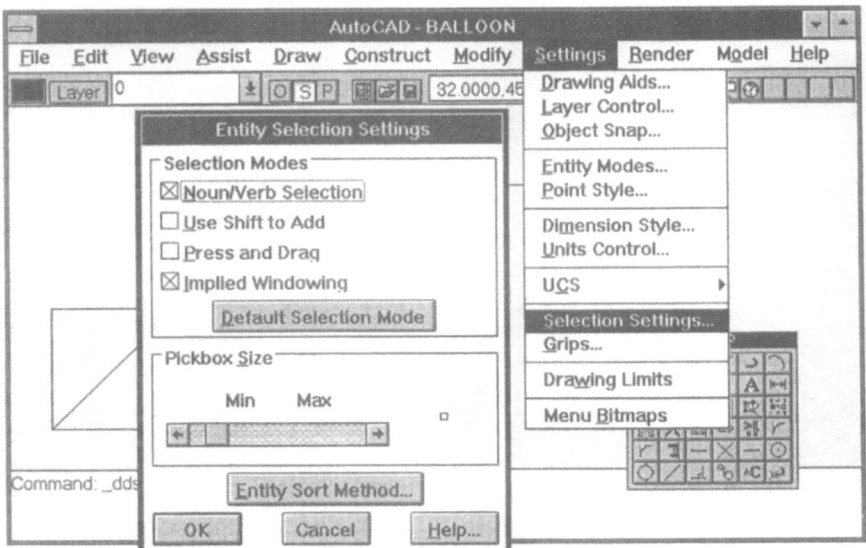

Figure 4.5 Enabling Noun/Verb Selection mode

Noun/Verb and Implied Windowing have an "x" in their boxes (ie are enabled) and click on the **OK** button. The default setting should have these enabled.

The way lines or other objects are selected for erasure is flexible. You can place the cursor on the item to be scratched and press the pick button, move to the next item and pick it and so on. You can select a whole group of entities using a window like in the ZOOM command or you can erase just the last object that was drawn. Indeed, the entity selection procedure can even use a combination of all these methods.

Erasing some lines

Firstly, you will erase two of the lines by simply picking them (Figure 4.6) and then picking the Erase button on the tool box. Where the cursor cross hairs meet there is a small box, called the pickbox. This box indicates that you are in the selection mode. Move to the lines and pick the points P1 (**11,27**) and P2 (**11,32**). When the line is successfully selected, little boxes, called grips, appear and the lines become ghosted. These give a visual acknowledgement of the selection. The grips can also be used in some advanced editing procedures. Now pick the **Eraser button** from the tool box. The command line will echo that 2 objects have been found and they will be erased.

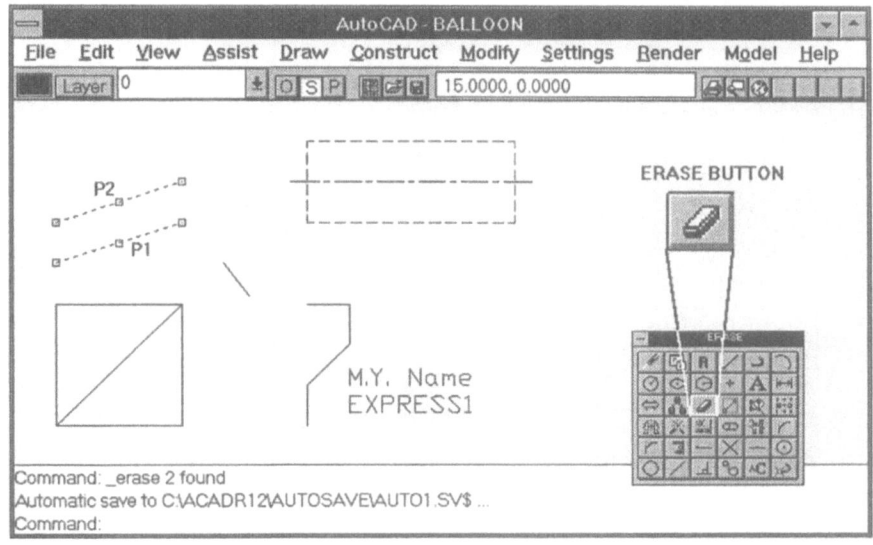

Figure 4.6 Erasing two lines

If you miss the object, it won't go dotty and the cursor will go into a windowing selection method. You can cancel the window by ˆC and resume picking. The next deletion demonstrates how to use windows to select objects.

Window selection

With window selection the order and relative position of the two corners is significant. For example, the window picked from W1 (**23,18**) to W2 (**64,43**) shown in Figure 4.7 finds the four lines, ghosts them and displays their grip points. Now pick the **ERASE** button and they disappear. Note that the text which is partially within the window is not selected. Only items that are fully within the window are selected. If any part of an entity is outside the window it will not be erased. For this mode of window selection, the first point picked must be to the left of the second as in this example.

OOPS! I didn't mean to rub that out

If unexpected items disappear from the display, first try a **REDRAW** and then type the **OOPS** command.

    If one item overlaps another and one of them is erased it looks like all the overlapped section has been deleted. Executing a **REDRAW** corrects this

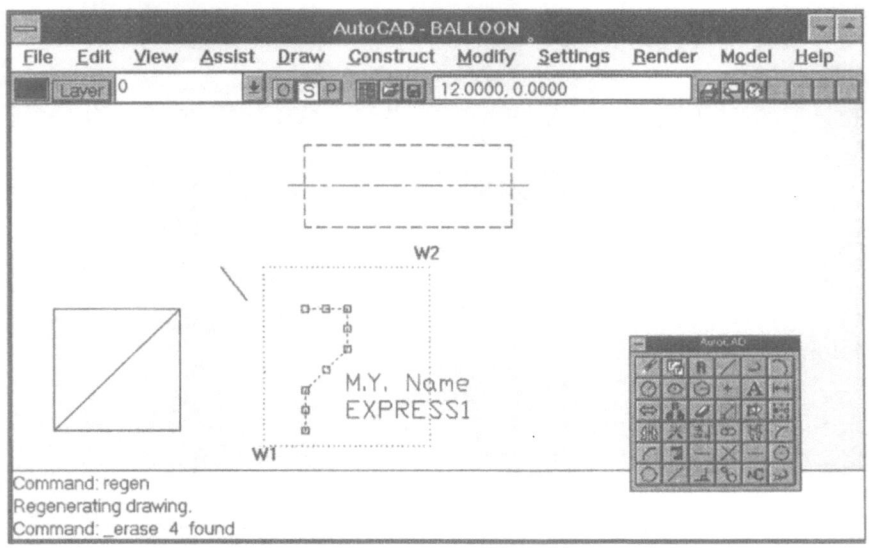

Figure 4.7 Erasing Window

display error. If the missing item doesn't reappear then it probably has been deleted. Type **OOPS** at the command line. This restores everything that was deleted by the most recent ERASE command execution.

    Command: **OOPS**

This restores the four lines that had just been erased.

    For more serious blunders you can step back in time by using the UNDO command. With UNDO you tell AutoCAD how many steps or command executions to go back. For example, the commands given below could be used to undo the last three operations.

    Command: **UNDO**
    Auto/Back/Control/End/Group/Mark/<Number>: **3**

HAZARD WARNING! This is a dangerous command as it can undo everything that you have done in the current editing session. If you then discover that you have undone too many commands the situation can be retrieved by executing a **REDO** immediately.

Command: **REDO**

Note that this command is more versatile than the "undo" which is available within the LINE command sequence. The general UNDO command cannot be executed transparently.

One of the best features of UNDO is creating marks. Before trying out some complicated manoeuvre you can set a mark. If things don't work out as planned you can use UNDO followed by Back to get back to the drawing as it was when you made the mark.

    Command: UNDO
    Auto/Back/Control/End/Group/Mark/<Number>: M
    . . .
    Do a series of AutoCAD commands.
    Command: UNDO
    Auto/Back/Control/End/Group/Mark/<Number>: B

Crossing windows and polygons

As mentioned above, the relative positions of the two corners of the window affect the behaviour. If the first corner is to the right of the second, the "Crossing" method is used. For example to re-erase the previous lines, pick the point W3 (**64,43**) and then drag the other corner to W4, (**23,18**), Figure 4.8. This selects the four lines as before but adds in the two lines of text. Now, pick the ERASE button and all six entities will disappear.

For the next couple of deletions we will use the Verb/Noun method. There is no particular significance in this choice. The selection methods used above can equally be employed after picking ERASE as before. Likewise, the selection methods used below could be used with Noun/Verb procedures.

Firstly, to get rid of the lines indicated in Figure 4.9, the regular window is no good. The standard window is always rectangular and parallel to the screen axes. What is needed here is a skew window or "WPolygon". Pick the ERASE button or type the command.

    Command: **ERASE**
    Select objects: **WPOLYGON** (or just WP)
    First polygon point: pick point **A**
    Undo <Endpoint of line>: pick point **B**
    Undo <Endpoint of line>: pick point **C**
    Undo <Endpoint of line>: pick point **D**
    Undo <Endpoint of line>: pick point **<ENTER>**

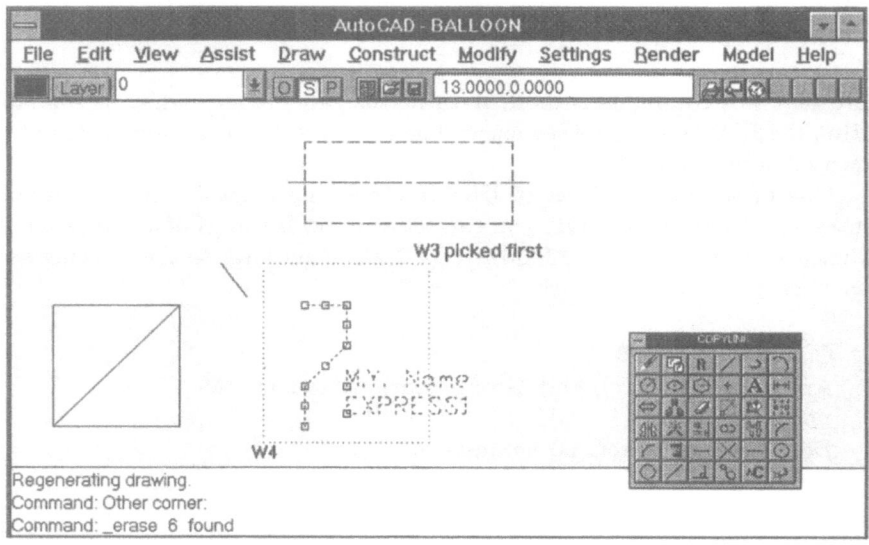

Figure 4.8 Erase Crossing

2 found.
Select objects: **<ENTER>**

This polygon works like the standard window. Only entities that are fully within the polygon are selected. The polygon can be any shape or size but lines must not cross. To tell AutoCAD that all the corners have been input press <ENTER> without picking a point. With the Verb/Noun procedure, ERASE prompts you to select the objects. At this prompt you can use any of the methods shown in the **Edit/Select** pull-down menu, Figure 4.10. The uppercase letters indicate the permitted truncations (except for the POINT option where you just pick points). When the window has been completed AutoCAD echos the number of entities found and ghosts them. It then prompts you to select more objects. When you are done, just press <ENTER> and all the ghosted objects will go.

The SELECT command is the explicit form for creating a selection set of entities. Again, you select the objects before deciding what to do with them. Here, we will use the "Fence" option to select four lines ghosted in Figure 4.11 and then delete them. The fence selects all objects that it crosses.

Command: **SELECT**

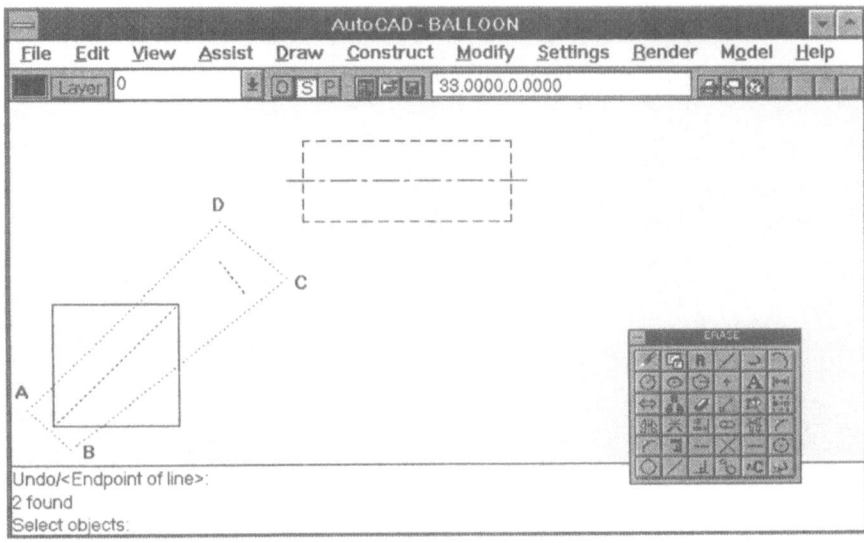

Figure 4.9 Erase Polygon

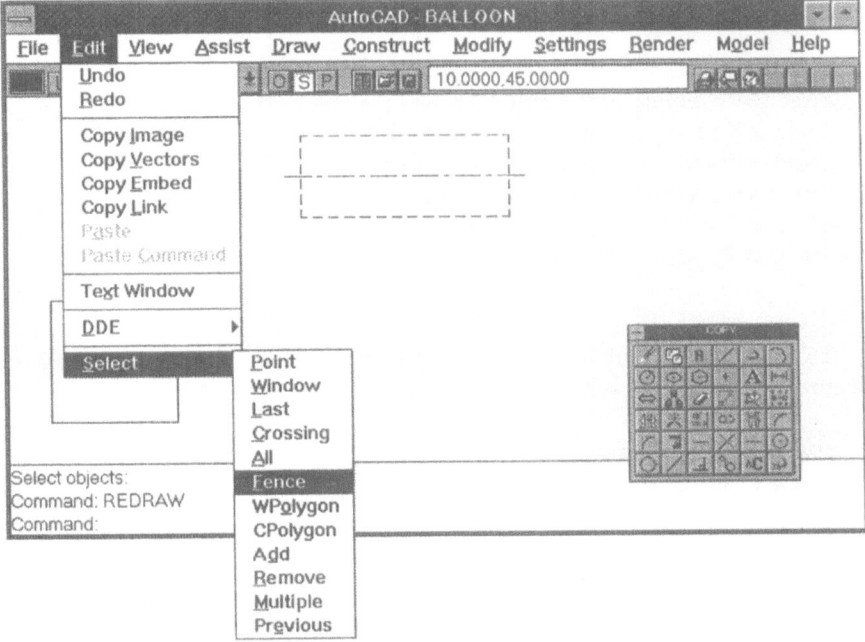

Figure 4.10 Select pull-down menu

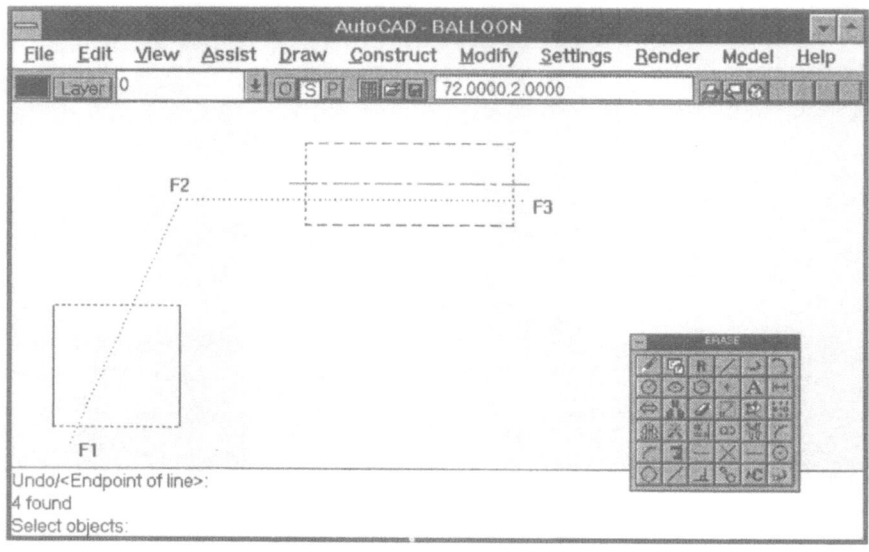

Figure 4.11 Fence selection method

Select objects: **Fence**
First fence point: pick point **F1**
Undo/<Endpoint of line>: pick point **F2**
Undo/<Endpoint of line>: pick point **F3**
Undo/<Endpoint of line>: pick point **<ENTER>**
4 found
Select objects: **<ENTER>**

To delete these four lines you need to execute the ERASE command and call up the "Previous" selection set, ie the one just done.

Command: **ERASE**
Select objects: **Previous**
4 found
Select objects: **<ENTER>**

The selection process used with the SELECT and ERASE commands is exactly the same for many other commands. Each object that has been ghosted becomes part of a "selection set". The selection set can have more objects added to it or have some removed.

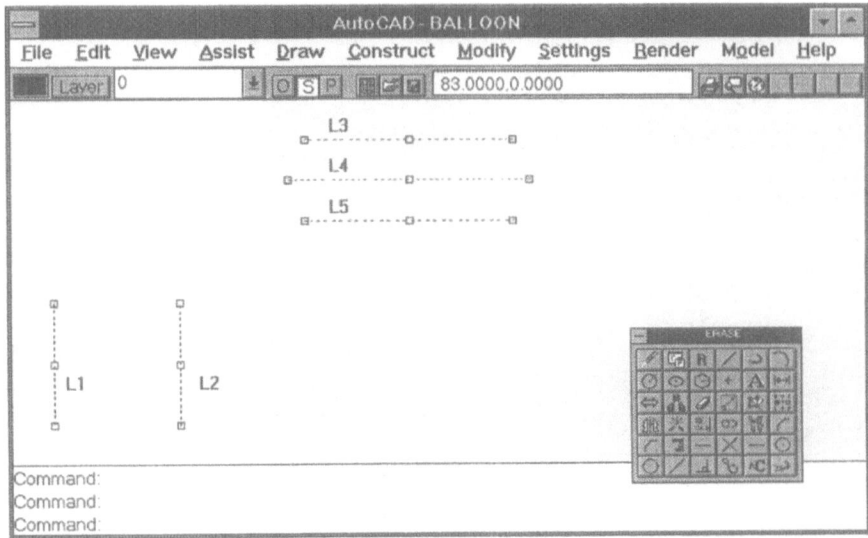

Figure 4.12 Erasing the last lines

Other options in the "Select" menu are Last, ALL, CPolygon, Add, Remove and Multiple. "Last" selects the most recently drawn object. "ALL" selects everything in the drawing including objects on invisible and frozen layers. This is usually done in conjunction with the "Remove" option, eg select everything except...

Typing or picking the keyword "Remove" changes the selection process from include to exclude mode. After picking "Remove" any objects selected are taken out of the selection set. "Add" goes back to the "include" mode. Try this out with the last five lines shown in Figure 4.12.

> Command: **ERASE**
> Select objects: pick lines **L1** to **L5**
> Select objects: **Remove**
> Remove objects: pick line **L1**
> 1 found, 1 removed.
> Remove objects: **<ENTER>**

That should have un-ghosted line L1 and saved it from deletion. However, the stay of execution is only temporary. As we need a clean sheet for the rest of the exercise delete it as follows.

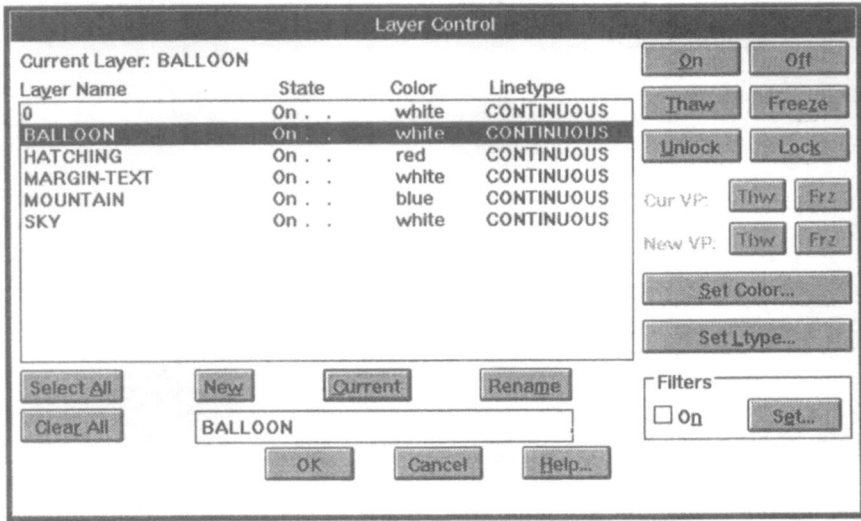

Figure 4.13 Layers for Balloon drawing

Command: **ERASE**
Select objects: pick line **L1**
1selected, 1 found.
Select objects: **<ENTER>**

To wind up, the CPolygon works in a similar way to WPolygon and Crossing.
It selects everything that is inside or crosses the polygon. Implied windowing
works as before. If no object is picked and the first point picked is to the left
of the second point then the standard "window" is invoked, otherwise it's like
"crossing".

The Balloon drawing environment

The drawing limits and available linetypes etc have been inherited from
the prototype drawing, EXPRESS1. That drawing contained only one layer,
namely the default 0 layer. In the following exercise you will need an additional five layers with appropriate color settings. To set up the layers shown
in Figure 4.13 pick **Settings/Layer Control** ... from the menu bar. Then
move the cursor arrow to the input box and type the layer names, "BAL-
LOON, HATCHING, MARGIN-TEXT, MOUNTAIN,SKY". Then pick the
New button.

To set the HATCHING color to red pick the line:

HATCHING On . . white CONTINUOUS

Then click **Set Color** followed by the **red** box from the color chart. Then pick **OK**. Finally, deselect the HATCHING layer by clicking it once more from the Layer Names list. Repeat this operation to set the layer, MOUNTAIN, to blue. Finally, select the line

BALLOON On.. white CONTINUOUS

and pick the **Current** button. This sets the BALLOON layer, ready to receive some circles. If the Current button appears grey and cannot be picked then check that only one layer is highlighted in the Layer Name list. If more than one layer is highlighted then click on their names to deselect them. When only **BALLOON...** is selected pick **Current**.

Creating circles, arcs and ellipses

Circles

Circles can be drawn in a variety of ways. The first circle will be specified by its centre point and radius while the second by centre and diameter. Pick the **Circle button** from the tool box (left-hand side of the second row).

```
Command: CIRCLE
3P/2P/TTR/<Center point>: 20,30
Diameter/<Radius>: 10
Command: <ENTER>
CIRCLE 3P/2P/TTR/<Center point>: @
```

This is equivalent to @0,0 from the last point, ie the centre of the first circle.

```
Diameter/<Radius>: Diameter
Diameter <20.0000>: 21
Command:
```

This gives two concentric circles (Figure 4.14). On some displays it might be difficult to distinguish between the two circles. Use ZOOM to check that they are really there.

All the methods of circle creation appear in the **Draw/Circle** pull-down menu. The two point circle is identified by picking the ends of one of its diameters. The three point circle allows some nice geometric constructions.

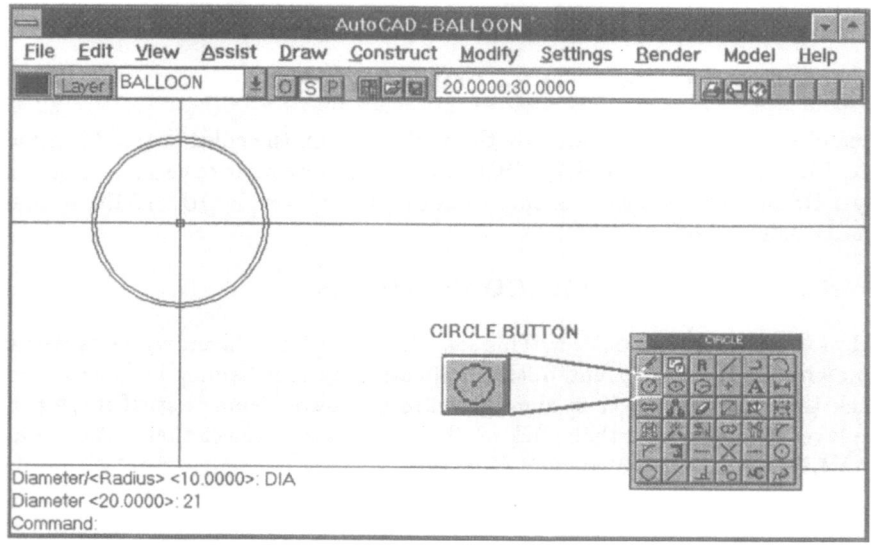

Figure 4.14 Concentric circles

The TTR allows you to draw a circle tangential to two objects and with an input radius. More circles will be covered in Chapters 7 and 8.

Many ways to draw an ARC

Before adding a cloud to the sky we must switch layers. The easiest way to do this in AutoCAD for Windows is to pick the "Down-pointing arrow" from the tool bar above the drawing area (next to where the name of the current layer is displayed). This displays a pull-down list of all the layers as shown in Figure 4.15. There is a scroll bar to the right of this pull-down for moving up and down long lists of layers. Pick **SKY** from the list and the current layer should change.

If you pick **Draw/Arc** from the menu bar you get a long list of arc creating options. At first this is a bit daunting with all the similar looking methods. There is also an Arc button in the tool-box.

To draw a cloud for the sky beside the balloon, you will need six arcs. The first arc will be a semi-circle. Since the angle within a semi-circle is 180 degrees the **Start End,Angle** (Figure 4.16) option is an appropriate choice. If you had to alter your units in Chapter 2, you will have discovered that angles are measured positive in the anti-clockwise (or counter-clockwise) direction. This

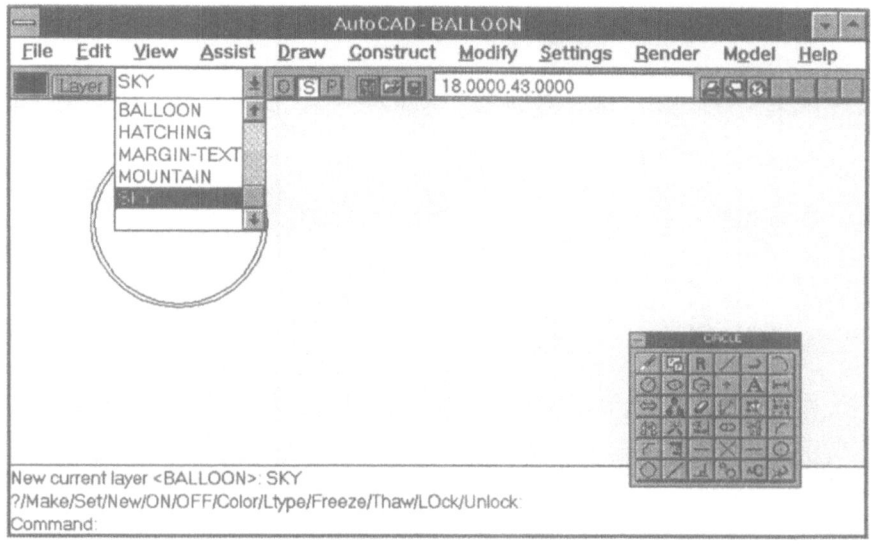

Figure 4.15 Switching the current layer

means that the start and end points should always be picked so that the arc joining them will go anti-clockwise.

    Command: **ARC** Center/<Start point>: **55,33**   (A1)
    End point: **47,33**   (A2)
    Angle/Direction/Radius/<Center point>: A Included angle: **180**

The second arc can be drawn using the **Start,End,Radius** with a radius of 3.

    Command: **ARC** Center/<Start point>: **48,34**   (B1)
    End point: **@5<265** hfill (B2)
    Angle/Direction/Radius/<Center point>: R Radius: **3**

Do the next two arcs with the **Start,Centre,End** approach.

    Command: **ARC** Center/<Start point>: **45,31**   (C1)
    Center/End/<Second point>: C Center: **45,28**   (C2)
    Angle/Length of chord/<End point>: DRAG **@5<326**   (C3)
    Command: **ARC** Center/<Start point>: **45,2**   (D1)
    Center/End/<Second point>: C Center: **50,32**   (D2)
    Angle/Length of chord/<End point>: DRAG **@15<307**   (D3)

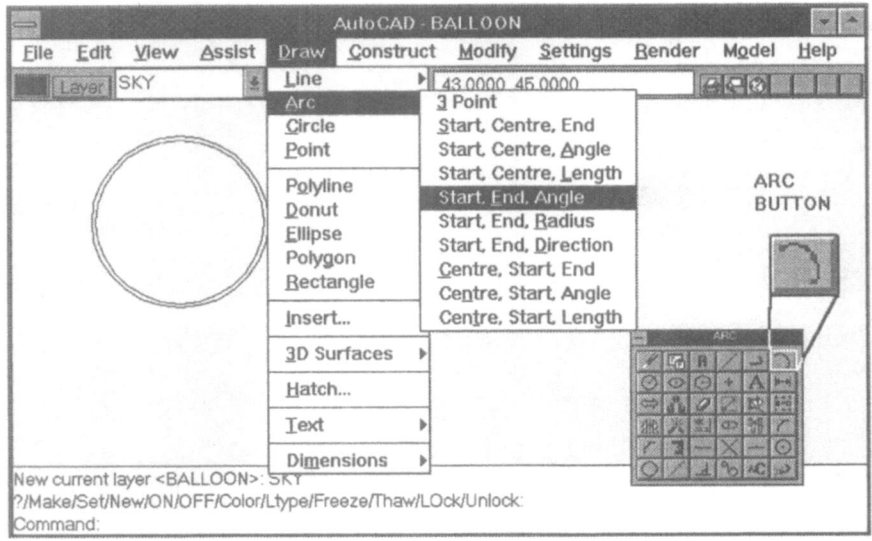

Figure 4.16 Switching the current layer

The angles used for the end points look a bit strange, but they are the result of dragging the arc until it looked right. The actual end points are not on the points C3 and D3, but are on the intersection between the arc whose radius is calculated from the start and centre points, and the line from the centre to the points C3 and D3. Figure 4.17 shows this procedure for arc D.

The remaining two curves to finish off the cloud can be drawn with **3 Point** arcs.

Command: **ARC** Center/<Start point>: **52,26**
Center/End/<Second point>: **57,25**
End point: DRAG **59,32**
Command: **ARC** Center/<Start point>: **59,30**
Center/End/<Second point>: **60,35**
End point: DRAG **54,34**

Rectangles and ellipses

You can make rectangles by drawing four connected LINEs, but this would be an assembly rather than a single entity. Thus to erase the rectangle formed with LINEs you would have to pick all four sides in the selection of objects.

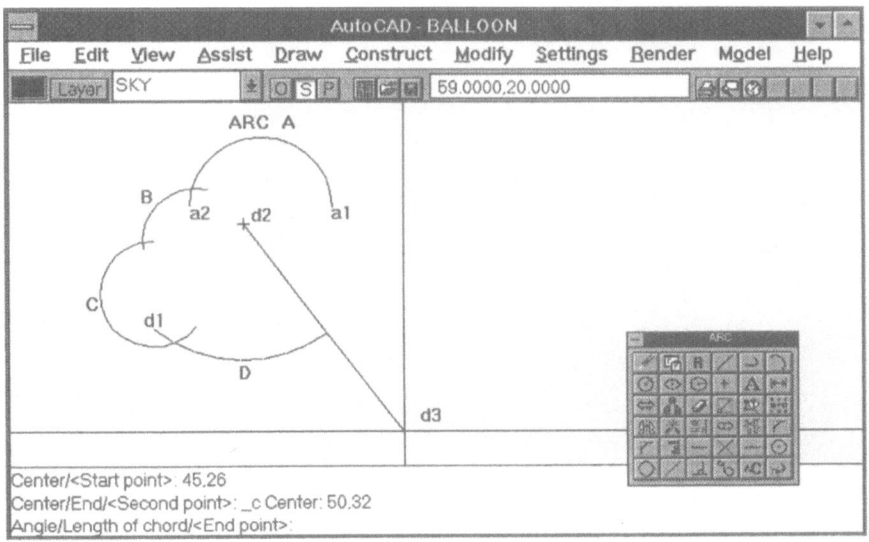

Figure 4.17 Dragging the ARC

The difference with PolyLINEs is that all the segments are considered as part of one entity. PLINE can be used in all the ways that LINE is used and a few more besides. Firstly, you can use it in place of LINE to draw the rectangle for the balloon's basket (Figure 4.18). Some special shapes use polylines in their construction eg rectangles, ellipses donuts and polygons.

You can draw the rectangular gondola beneath the balloon by picking **Draw/Rectangle**. This constructs a rectangle using a polyline. For example, the ABCD rectangle could have been drawn as follows:

```
        Command: _rectang
        First corner: 15,10                                              (A)
        Other corner: @10,5                                              (C)
```

Exactly the same gondola could have been drawn using the PLINE command but you would have to pick each of the points ABCD. This command is fully described in the section on Wide Lines later in this chapter. If your rectangle appears with very thick lines then your polyline width has probably got a non-zero value. Skip forward to the section on Wide Lines and set this to zero, erase the thick rectangle and draw it again.

Note that this polyline rectangle is one single entity. If you try to **ERASE** it just pick one point on the rectangle and the whole thing will be selected. Use ^C to avoid rubbing it out or **OOPS** to bring it back.

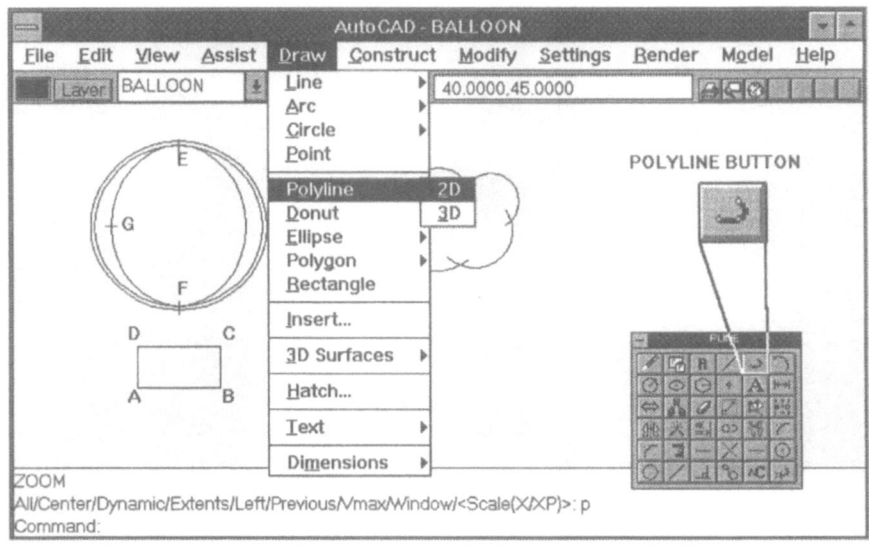

Figure 4.18 Basket and ellipse

To draw the ellipses on the smaller circle of the balloon pick **Draw/Ellipse** from the menu bar. You will be prompted to give the two end points of one of the axes and one end of the other axis. In this example, you will use object snap QUADrant to locate the top and bottom of the inner circle for the major axis.

Command: **ELLIPSE**
\<Axis endpoint 1>/Center: **QUADRANT** of **pick point E**   (20,40)
Axis endpoint 2: **QUADRANT** of **pick point F**   (20,20)
\<Other axis distance>/Rotation: **12,30**   (G)

Repeat this procedure for two more ellipses to complete the balloon (see Figure 4.19 below).

Command: **ELLIPSE**
\<Axis endpoint 1>/Center: **QUADRANT** of **pick point E**   (20,40)
Axis endpoint 2: **QUADRANT** of **pick point F**   (20,20)
\<Other axis distance>/Rotation: **15,30**
Command: **\<ENTER>**
ELLIPSE
\<Axis endpoint 1>/Center: **QUADRANT** of **pick point E** (20,40)

Axis endpoint 2: **QUADRANT** of **pick point F** (20,20)
<Other axis distance>/Rotation: **18,30**

To connect the basket to the balloon, some cables must be provided. Use SNAP to locate the end points on the basket and object snap TANGENT on the ellipses. Pick **Draw/Line/1 Segment.**

Command: **LINE** From point: **15,15**
To point: **TANGENT** To **10,30**   (inner circle)
To point:
Command: **<ENTER>**
LINE From point: **17,15**
To point: **TANGENT** To **12,30**   (first ellipse)

Make sure that you pick the ellipse near the point (12,30). If you pick it at other locations, you may get a message saying that no tangent exists. This is because the ellipse is constructed using a series of circular arcs. Some arcs may not have a valid tangent.

To point: **<ENTER>**
Command: **<ENTER>**
LINE From point: **19,15**
To point: **TANGENT** To **16,30**   (second ellipse)
To point: **<ENTER>**

And for the cables on the right hand side:

Command: **<ENTER>**
LINE From point: **21,15**
To point: **TANGENT** To **25,30**   (second ellipse)
To point: **<ENTER>**
Command: **<ENTER>**
LINE From point: **23,15**
To point: **TANGENT** To **28,30**   (first ellipse)
To point: **<ENTER>**
Command: **<ENTER>**
LINE From point: **25,15**
To point: **TANGENT** To **30,30**   (inner circle)
To point: **<ENTER>**

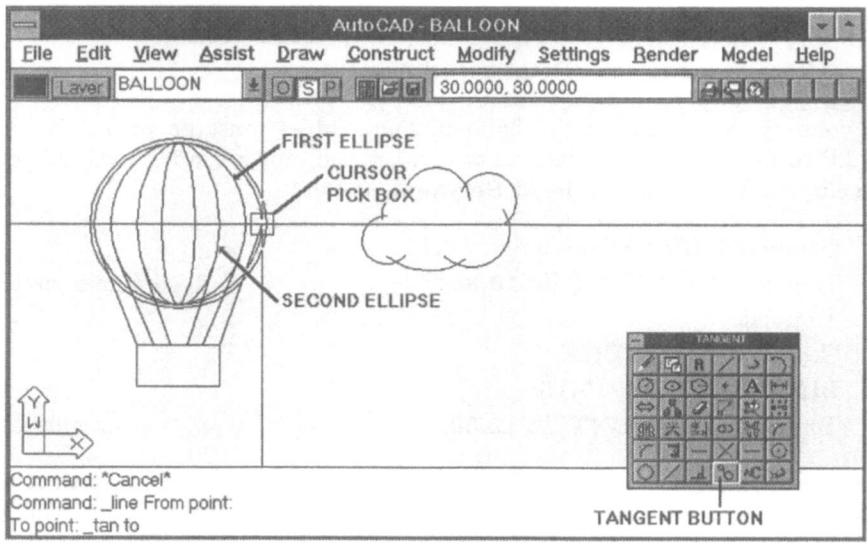

Figure 4.19 More ellipses and cables

Adding text

Textual annotations are an important part of any engineering drawing. The annotations may include information about the objects on the drawing or about the drawing itself. AutoCAD allows you to insert text on a drawing giving you full control over how this should be done and how the text should look.

Already in the drawing EXPRESS1 you have added your name using the DTEXT command. In that example you picked the start point of the text and its height and rotation. To repeat a similar operation, switch to the layer MARGIN-TEXT and try the following. You can switch layers by picking the "Down-pointing arrow", beside the layer name, from the tool bar and clicking MARGIN-TEXT. Intrepid typists could use the keyboard!

Command: **LAYER**
?/Make/Set/...: **S**
New current layer <0>: **MARGIN-TEXT**
?/Make/Set/...: **<ENTER>**

The DTEXT, or Dynamic TEXT, button is denoted by a large letter "A" on the floating tool box (Figure 4.20).

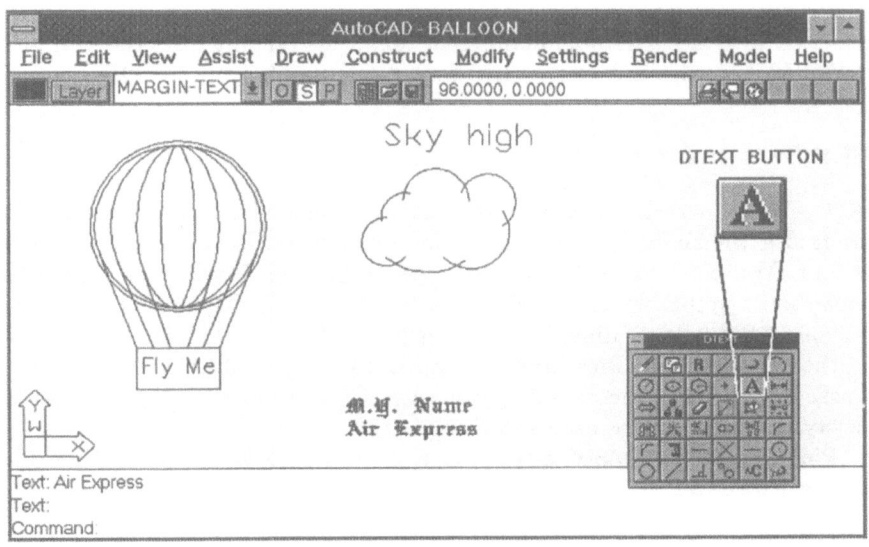

Figure 4.20 Adding Text

Command: **DTEXT** or pick the button
Justify/Style <Start point>: **45,40**
Height <2.00>: **2.5**
Rotation angle <0>: **ENTER>**
Text: **Sky high <ENTER>**
Text: <ENTER>   DTEXT expects multiple lines of text
Command:

This gives left justified text, ie the first letter is lined up with the start point.

The most common mistake people make when entering text is not paying attention to the prompt line. As most text is inserted with an angle of 0 degrees it is easy to forget that you have to re-input the angle every time. This type of error is more likely to crop up if you are inputting a lot of text and paying more attention to the difficult operation of typing than to the screen.

Of the other options for inserting text, "Justify" is the next most useful. It can be used to centre the characters on a point or to right justify the text. This is particularly good for making up posters and signs. Remember, to make the most impact with text, it must look good. The default type of lettering available is called the SIMPLEX font which is simple, fast and memory efficient.

AutoCAD provides a number of different text fonts ranging from the simplicity of SIMPLEX to the extremely complicated and flamboyant GOTHIC.

Text styles

AutoCAD has a rather confusing arrangement of text fonts and styles. You cannot use the fonts by themselves but must define a text style based on the font. You can create a new text style by picking **Set Style...** from the **Draw/Text** pull-down menu (Figure 4.21). AutoCAD displays all its fonts in the "Select Text Font" dialogue box. Click the **Next** or **Previous** buttons to find the one you want. Pick **Gothic English** and then pick **Ok**. You will then be asked to define some default parameters. The sequence below defines two new styles which will be used shortly.

Pick **Draw/Text/Set Style** ... from the pull-down menu. Then pick **Gothic English** from the list in Figure 4.21. The Command line will echo:

Command: STYLE Text style name (or ?) <STANDARD>: GOTHICE
New style.
Font file <txt>: GOTHICE   (note the spelling)

At this point AutoCAD requests some parameters which can be used to customize the text style. Here we just accept the defaults.

Height <0.0000>: <ENTER>
Width factor <1.00>: <ENTER>
Obliquing angle <0>: <ENTER>
Backwards? <N>: <ENTER>
Upside-down? <N>: <ENTER>
Vertical? <N>: <ENTER>
GOTHICE is now the current text style.

Now do the same to create a simplex style. This time pick the "Roman Simplex" font.

Command: <ENTER> (Pressing <ENTER> repeats the last command)
STYLE Text style name (or ?) <GOTHICE>: B-SIMPLEX
New style.
Font file <txt>: SIMPLEX or pick it and then "OK" in the dialogue box
Height <0.0000>: <ENTER>
Width factor <1.00>: <ENTER>
Obliquing angle <0>: <ENTER>
Backwards? <N>: <ENTER>
Upside-down? <N>: <ENTER>
Vertical? <N>: <ENTER>

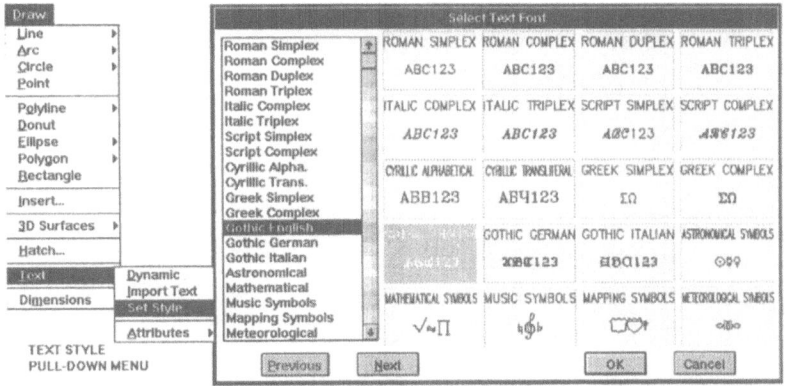

Figure 4.21 Text fonts

B-SIMPLEX is now the current text style.

All new text will now be drawn in the new style. The height of 0.0000 specified for the default simply means that the default will vary. An obliquing angle could be used to form an *italic* style and backwards text produces mirror writing.

Centered text

Use the "Justify" option to position a message centrally in the balloon's basket.

> Command: **DTEXT** or pick the "A" button fromn the tool box
> Justify/Style <Start point>: **J**
> Align/Fit/Center/...: **C**

You must use the abbreviated form for the option. If you type "center", AutoCAD will assume you want to use Object Snap and tries to find the centre of an arc or circle.

> Center point: **20,12**   (middle of basket)
> Height <2.5>: **1.8**
> Rotation angle <0>: **<ENTER>**
> Text: **Fly ME**
> Text: **<ENTER>**

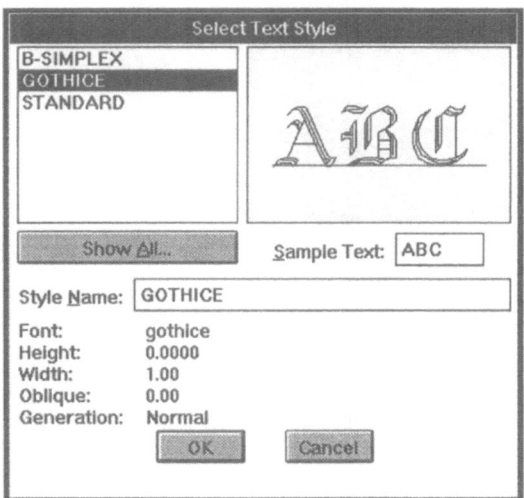

Figure 4.22 Text Style Dialogue Box

Of the other justifying options, inserting text using the "Middle" option is similar to centred but text is balanced about the point horizontally and vertically. With "C" the text is balanced horizontally. With "Aligned" text you give the start and end points and AutoCAD calculates the height so that your text just fits. The rotation angle is defined by the two end points. "Fit" is similar to aligned but you can specify a text height and AutoCAD calculates the text width. The final option, "Right", allows you to input the end point, height and rotation of the text and AutoCAD works out the start point. Finally, the abbreviations, TL ... BR indicate the top left or bottom right corners of the text.

You can change the text style to GOTHICE by typing "S" as the reply to the "Justify/Style..." prompt. This will only show the styles that have been defined in the current drawing (Figure 4.22).

> Command: **DTEXT**
> Justify/Style <Start point>: **S**
> Style name (or ?) <B-SIMPLEX>: pick **GOTHICE** from the dialogue box.
> Justify/Style <Start point>: **40,7**
> Height <2.5>: **1.8**
> Rotation angle <0>: **<ENTER>**
> Text: **M.Y. Name <ENTER>**
> Text: **Air Express <ENTER>**

Text: **<ENTER>**

This procedure makes it easy to enter multiple lines of text. AutoCAD automatically calculates the line spacing and uses all the responses given to the prompts at the first line. You can input multiple lines of centred or right justified text.

On serious drawings it is important to keep control of the text heights that are used. Too many different heights and/or fonts spoil the appearance of the drawing. Furthermore, it is vital that all text be legible on the final plot or print out. Thus, the text heights should be chosen carefully and with a view to the final legibility.

On large drawings the time taken to regenerate all the text can be considerable. While editing, you can speed things up by using the simple fonts, TXT and SIMPLEX, and changing to the fancy ones at the end. You can also put all the text on suitable layers and freeze them until the end. The disadvantage of the latter approach is that while the text is invisible, you may draw something on top of it. An alternative approach is to turn on the QTEXT or quick text mode. When QTEXT is on, all the text is displayed as rectangles. With this you at least know where the text is. **Quick Text** is on the Drawing Aids dialogue box and can be found by picking **Settings/Drawing Aids...** from the menu bar. Then click the box beside "Quick Text" followed by **OK**. The text display will not change until the next REGEN command is issued (Figure 4.23). You can also do this by typing the command.

Command: **QTEXT** ON/OFF <OFF>: **ON**
Command: **REGEN**   (to see the effect)

Turn **QTEXT** back **OFF** and draw the margin ABCD shown in Figure 4.23.

Command: **QTEXT**
ON/OFF <ON>: **OFF**
Command: **REGEN**

Wide lines

Up to this point, we haven't been too concerned with the width of the lines we were drawing. In fact everything we have drawn has had a notional line width of zero. AutoCAD uses zero width to mean that no matter how much you magnify the line it always appears the same width. Similarly, it doesn't matter how far you zoom out from an object, the width doesn't change.

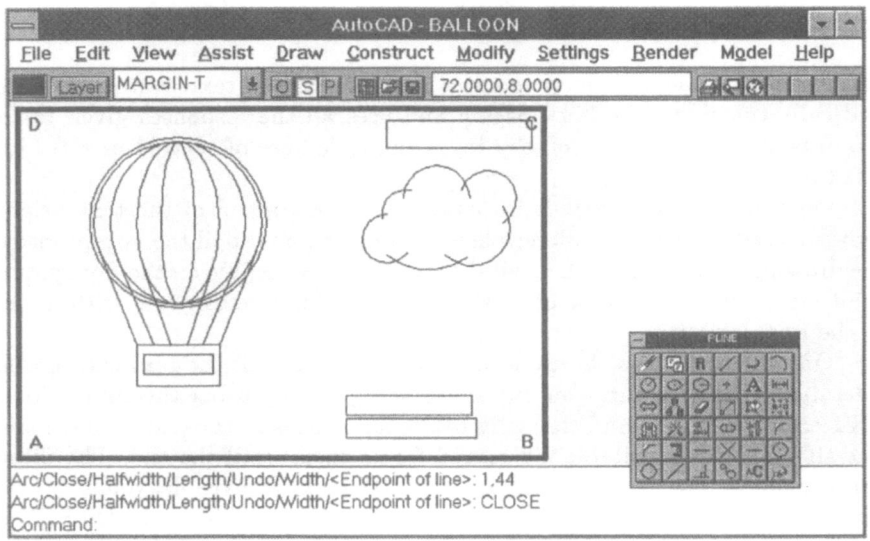

Figure 4.23 Quick text mode in the frame

There are many times when you might want to draw a line with a specified width: printed circuit board drawings, for example. In this section you will cover the methods of assigning widths and using them in the balloon drawing artwork. The same commands and methods can be used in any type of drawing where wide lines are required.

WARNING! Assigning width is mildly dangerous for plotted output.

Assigning non-zero widths should only be used where the width serves a particular purpose. If you have a drawing convention that, say, all outlines are to be drawn in 0.5mm line width then it might be better to assign a particular color to the layer containing the outlines. Then, at plot time, you can specify that the outline color will be drawn using a 0.5mm pen. If you assign a width on the AutoCAD drawing itself and plot the drawing at a scale of 1:10 the width will be plotted at the reduced width.

The best way of generating wide lines is to use the PLINE command. With polylines it is possible to assign a constant width or a varying width to the line. You can also draw wide arcs with PLINE. To draw a frame for the balloon drawing (Figure 4.23) use **Draw/Polyline/2D** from the menu bar and set a width of 0.5 units. Make sure ORTHO is on and SNAP is set to 1 and that the current layer is MARGIN-TEXT.

Command: **PLINE**
From point: **35,1**   (A)
Current line-width is 0.0000
Arc/Close/.../Width/<Endpoint of line>: **Width**
Starting width <0.0000>: 0.5
Ending width <0.5000>: **<ENTER>**
Arc/Close/.../<Endpoint of line>: **1,1**   (A)
Arc/Close/.../Width/<Endpoint of line>: **64,1**   (B)
Arc/Close/.../Width/<Endpoint of line>: **64,44**   (C)

No wide line appears on the screen until the third point is picked because AutoCAD must work out the mitre detail at the second point. This mitre depends on the angle between the first and second lines.

Arc/Close/.../Width/<Endpoint of line>: **1,44**   (D)
Arc/Close/.../Width/<Endpoint of line>: **Close**   (A)

If your polyline is drawn in outline only, then you should turn the Solid Fill on from the Drawing Aids dialogue box (Settings/Drawing Aids...) and REGENerate the drawing or type the command.

Command: **FILL**
ON/OFF <current>: **ON**
Command: **REGEN**

Filling in wide lines and solid objects can slow down AutoCAD's responses. Turning FILL off will speed up the REDRAW and REGEN commands.

In the following sequence of operations you will add some mountain scenery to the picture (Figure 4.24). Before putting in these mounatains we must switch to the layer, MOUNTAIN. Click the Layer pull-down icon (the down- pointing arrow) from the tool-bar and then click **MOUNTAIN** from the resulting list. The Command line will echo the following.

Command: LAYER
?/Make/Set/...: S
New current layer <0>: MOUNTAIN
?/Make/Set/...:

The first PLINE will have a series of segments with two different but constant widths. The second will have varying widths, while a third will combine varying width with arcs. The listing of the commands and prompts looks long, but most of the work is done by AutoCAD. If you pick any wrong points or get the widths mixed up, use the "Undo" option to step backwards along the polyline.

Command: **PLINE**
From point: **29,4**   (F)
Current line-width is 0.5000
Arc/Close/. . ./Width/<Endpoint of line>: **W**
Starting width <0.0000>: **0.1**
Ending width <0.1000>: **<ENTER>**
Arc/Close/. . ./Width/<Endpoint of line>: **@6,10**   (G)
Arc/Close/. . ./Width/<Endpoint of line>: **@4,−2**   (H)
Arc/Close/. . ./Width/<Endpoint of line>: **@4,5**   (J)

Now change the width to 0.75 units.

Arc/Close/. . ./Width/<Endpoint of line>: **W**
Starting width <0.1000>: **0.75**
Ending width <0.7500>: **<ENTER>**
Arc/Close/. . ./Width/<Endpoint of line>: **@4,−5**   (K)
Arc/Close/. . ./Width/<Endpoint of line>: **@8<60**   (L)

Now change over to drawing an arc.

Arc/Close/. . ./Width/<Endpoint of line>: **Arc**
Angle/CEnter/. . ./Width/<Endpoint of arc>: **@13,3**   (M)
Angle/CEnter/. . ./Width/<Endpoint of arc>: **<ENTER>**

The arcs in a polyline are always drawn tangential to the previous PLINE
segment. This ensures a smooth transition between the straight lines and the
curve.

The mountains on the other side of the balloon will be made up of varying
width polylines.

Command: **PLINE**
From point: **2,21**   (N)
Current line-width is 0.7500
Arc/Close/. . ./Width/<Endpoint of line>: **W**
Starting width <0.7500>: **0**
Ending width <0.0000>: **1**
Arc/Close/. . ./Width/<Endpoint of line>: **9,12**   (P)
Arc/Close/. . ./Width/<Endpoint of line>: **W**
Starting width <1.0000>: **<ENTER>**
Ending width <1.0000>: **3**
Arc/Close/. . ./Width/<Endpoint of line>: **20,3**   (Q)

And now to reduce the half-width back to zero and connect up with the first
PLINE.

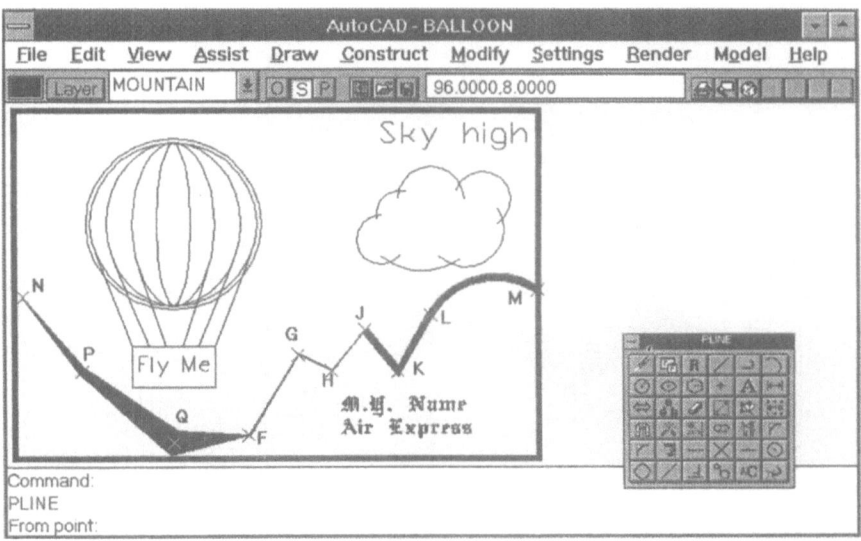

Figure 4.24 Varying PLINE's width

Arc/Close/. . ./Width/<Endpoint of line>: **W**
Starting width <3.0000>: **<ENTER>**
Ending width <3.0000>: **0**
Arc/Close/. . ./Width/<Endpoint of line>: **29,4**   (F)
Arc/Close/. . ./Width/<Endpoint of line>: **<ENTER>**

The final polyline will be used to draw the silhouetted bird in flight (Figure 4.25). Every self-respecting sky should have one. This is achieved by starting out with a straight line segment with a width going from zero to 0.4. This is then blended with an arc whose width increases from 0.4 to 0.8 followed by a short line segment. The reverse procedure gives the other wing. As this is an intricate manoeuvre, it is prudent to ZOOM in first.

Command: **ZOOM**
All/Center/. . ./Window/<Scale(X/XP)>: **W**
First corner: **29,16**
Other corner: **42,24**
Command: **PLINE**
From point: **30,18**   (A)
Current line-width is 0.0000

Set up varying width.

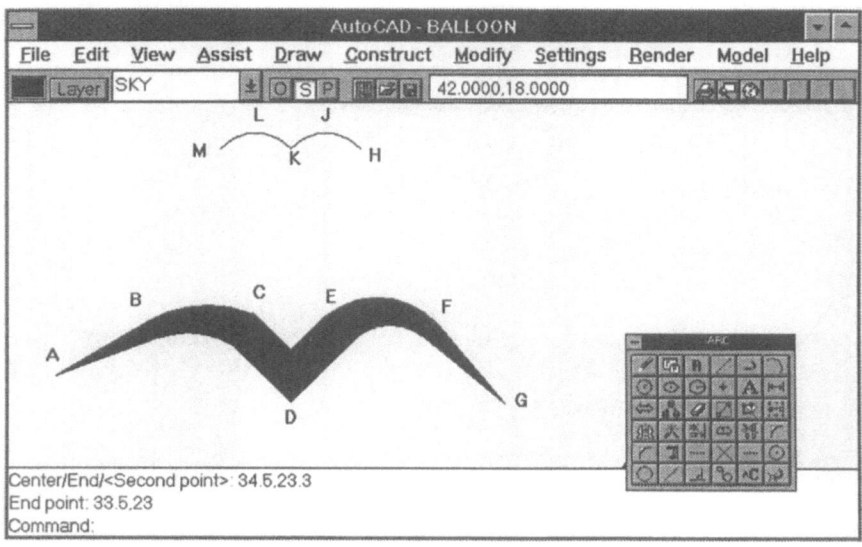

Figure 4.25 Silhouetted birds

Arc/Close/.../Width/<Endpoint of line>: **W**
Starting width <0.0000>: **<ENTER>**
Ending width <0.0000>: **0.4**
Arc/Close/.../Width/<Endpoint of line>: **@2,1**   (B)
Arc/Close/.../Width/<Endpoint of line>: **W**
Starting width <0.4000>: **<ENTER>**
Ending width <0.4000>: **0.8**

Change over to drawing arcs.

Arc/Close/.../Width/<Endpoint of line>: **ARC**
Angle/CEnter/.../Width/<Endpoint of arc>: **@2,0**   (C)

Change back to straight lines.

Angle/CEnter/.../Width/<Endpoint of arc>: **Line**
Arc/Close/.../Width/<Endpoint of line>: **@1,−1**   (D)
Arc/Close/.../Width/<Endpoint of line>: **@1,1**   (E)

And draw the second half of the bird.

Arc/Close/.../Width/<Endpoint of line>: **W**
Starting half-width <0.8000>: **<ENTER>**

Ending half-width <0.8000>: **0.4**
Arc/Close/.../Width/<Endpoint of line>: **ARC**
Angle/CEnter/.../Width/<Endpoint of arc>: **@2,0**   (F)
Angle/CEnter/.../Width/<Endpoint of arc>: **Line**
Arc/Close/.../Width/<Endpoint of line>: **W**
Starting width <0.4000>: **<ENTER>**
Ending width <0.4000>: **0**
Arc/Close/.../Length/Undo/Width/<Endpoint of line>: **L**
Length of line: **2.24**   (G)
Arc/Close/.../Width/<Endpoint of line>: **<ENTER>**

The second bird in the distance is just two normal three-point ARCs. Pick
Draw/Arc/3 Point from the menu bar.

Command: **ARC** Center/<Start point>: **36.5,23**   (H)
Center/End/<Second point>: **35.5,23.3**   (J)
End point: DRAG **35,23**   (K)
Command: **ARC**
Center/<Start point>: **@**   (K)
Center/End/<Second point>: **34.5,23.3**   (L)
End point: DRAG **33.5,23**   (M)

Aerial view

It can sometimes become a bit annoying to have to zoom in and out on a
drawing repeatedly to locate the next bit of drafting. This is particularly so
on large drawings. AutoCAD for Windows provides a facility called "Aerial
View" which helps to navigate through drawings without repeated zooms.
The Aerial View button looks like a miniature compass and is located towards
the right of the tool-bar (Figure 4.26).

Click this button and the Aerial View Window appears. The **Zoom** command should be highlighted. The window shows the complete drawing with a
box indicating the current zoom window. To select a new zoom window pick
the lower left corner and then the other corner, just above the "Fly Me" text
as shown.

The graphics display will be updated when the second point of the window
is picked. You can move the Aerial View window around the screen like any
other Windows program. You can move it out of the way by picking the title
bar and dragging it around. To get back to the drawing screen, just move the
cursor to the drawing area and click. The Aerial View can be closed by picking
control box in the top left corner followed by Exit. There will be more on
Aerial View later.

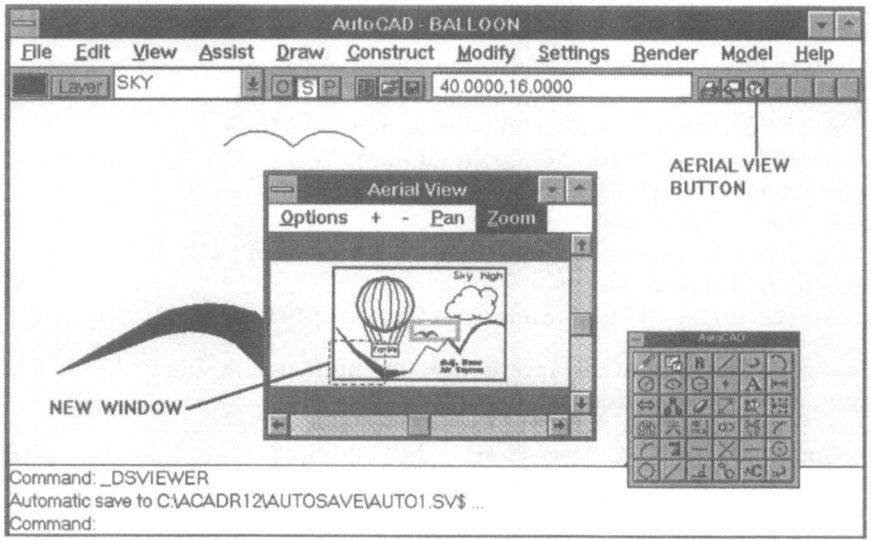

Figure 4.26 Aerial view of balloon

Solid objects

AutoCAD's SOLID command is used to create filled polygons. Its usage is a
little tricky. Here you will draw a little house on the hillside (Figure 4.27). This
is made up of a solid triangle on top of a rectangle. The Aerial View operation
shown in Figure 4.26 should display the bottom left corner of the drawing as
shown in Figure 4.27. If your screen doesn't match the new figure or if you're
not sure, you could execute the Zoom command.

> Command: **ZOOM**   (This command sequence is optional)
> All/Center/.../Window/<Scale(X/XP)>: **W**
> First corner: **0,0**
> Other corner: **26,16**

To draw the triangular roof is not too difficult. Type **SOLID** at the command
line and then pick the three points.

> Command: **SOLID** First point: **4,8**   (A)
> Second point: **10,8**   (B)
> Third point: **7,10**   (C)
> Fourth point: **<ENTER>**
> Third point: **<ENTER>**

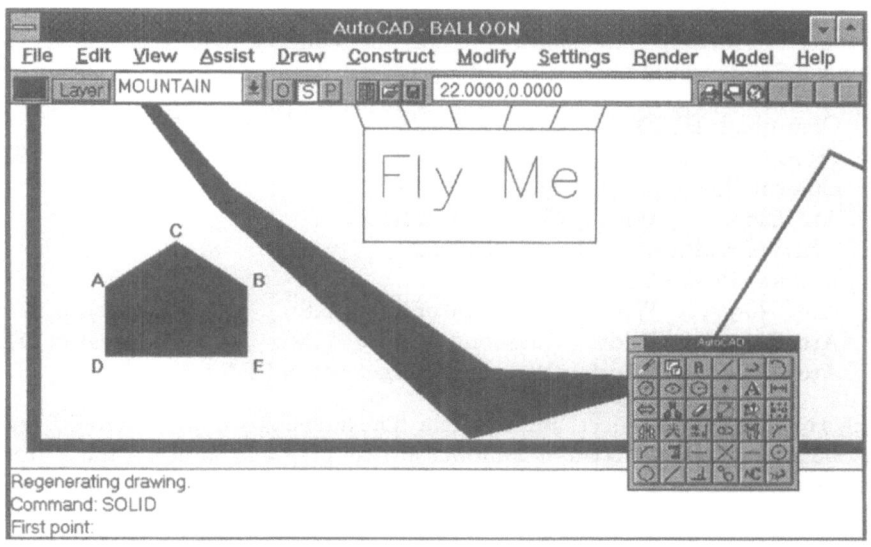

Figure 4.27 A solid house

Since there is no fourth point you simply hit **<ENTER>** or the space bar. The extra "Third point:" is to allow you to draw another solid using the last two points as the first and second. Pressing **<ENTER>** a second time exits from the solid command. For triangles the order of picking the points is not important. However, the order is crucial for four-sided objects. To draw the rest of the house, pick the points in the order given below.

Command: **SOLID** First point: **4,8**	(A)
Second point: **10,8**	(B)
Third point: **4,5**	(D)
Fourth point **10,5**	(E)
Third point: **<ENTER>**	

If you put in the points in the wrong order, you will get a bow-tie effect instead of a rectangle. You could have drawn the whole house and roof in one SOLID command sequence by picking the points in the order DEABC and then pressing <ENTER> twice. However, the chances of success are increased if you simplify the object and draw each triangle and rectangle separately.

Many users have difficulty with the SOLID command. If you do, then do not despair, for PLINE can do everything that SOLID can do. Another difficulty is possible confusion with the 3D solids provided by AutoCAD's

Advanced Modelling Extensions. The polyline sequence to do the same as the previous Solids would be:

Command: **PLINE**
From point: **7,10**   (C)
Current line-width is 0.0000
Arc/Close/.../Width/<Endpoint of line>: **W**
Starting width <0.0000>: **<ENTER>**
Ending width <0.0000>: **4**
Arc/Close/.../Width/<Endpoint of line>: **7,8**   (Midpoint of AB)
Arc/Close/.../Width/<Endpoint of line>: **7,5**   (Midpoint of DE)
Arc/Close/.../Width/<Endpoint of line>: **<ENTER>**

Here's one final solid object, a filled circle. The amusingly named DONUT (can also be spelt "DOUGHNUT") is in fact a little program based on the PLINE Arc. It can be found on the DRAW screen menu. As the name suggests you can use it to draw fat circles. If the inside diameter of the doughnut is zero then you have a filled circle. To draw the sun in Figure 4.31 you will first have to pick **View/Zoom All** from the menu bar and change the color to yellow.

Command: **ZOOM**
All/Center/.../<Scale(X/XP)>: **A**
Command: **COLOR**   (Or pick the color box from the tool-bar)
New entity color <BYLAYER>: **YELLOW** (Or pick from dialogue box)

Draw the sun. Pick **Draw/Donut** from the menu bar.

Command: **DONUT**
Inside diameter <0.5000>: **0**
Outside diameter <1.0000>: **6**
Center of doughnut: **37,40**
Center of doughnut: **<ENTER>**
Command: **COLOR**
New entity color <2 (yellow)>: **BYLAYER**

This means that the colors of new entities will be the same as their layer's default color. This was set up using the LAYER command.

Shading with patterns

Many types of architectural and engineering drawings use standard hatching patterns to indicate such things as cross sections and material type. AutoCAD contains a good library of hatch patterns conforming to ANSI (American National Standards Institute) norms.

For HATCH to work correctly, the area to shaded should be within a closed boundary. Furthermore, the entities, lines, arcs etc, making up the perimeter must intersect at their end points. If they don't, or if there are any protruding ends, the results may be incorrect with some strange shading.

It is always a good policy to make a small test box for hatching so that the effectiveness of the pattern scale can be assessed before hatching a larger area.

```
Command: LAYER
?/Make/Set/...: S
New current layer <0>: HATCHING
?/Make/Set/...: <ENTER>
Command: LINE
From point: 34,9
To point: @5,0
To point: @0,−5
To point: @−5,0
To point: Close
Command: SAVE
File name <BALLOON>: <ENTER>
```

HAZARD WARNING! The HATCH command is dangerous. Always save your drawing before executing this command.

The procedure for adding patterns to a drawing is controlled through the command **BHATCH** or Boundary HATCH. It is executed by picking **Draw/Hatch...** from the menu bar.

Command: **BHATCH**

This brings up the Boundary Hatch dialogue box shown in Figure 4.28. You will then have to select a hatch pattern followed by picking the object defining the boundary of the hatch area.

To select the hatch pattern pick the **Hatch Options** button. The Hatch Optinos dialogue box should now appear (Figure 4.29). Make sure that the Pattern Type, "Stored Hatch Pattern" is active. There should be large dot inside the button as shown in Figure 4.29. If the dot is in the User-Defined button simply pick the Stored Hatch button to change the setting.

Then pick **Pattern...** from the Hatch Options dialogue box (Figure 4.29). This brings up a window full of patterns. The one we want is called "Stars". As the patterns are stored on a series of windows and in alphabetical order you need to pick **Next** or type "N" three times to get to the one shown in Figure 4.30. If you press "Next" too many times then "Previous" allows you to go back.

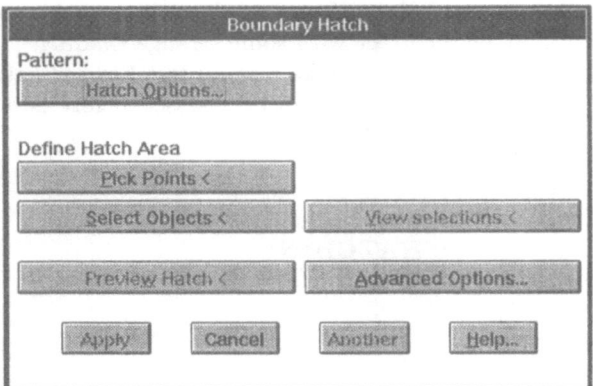

Figure 4.28 Boundary Hatch dialogue box

Figure 4.29 Hatch Options dialogue box

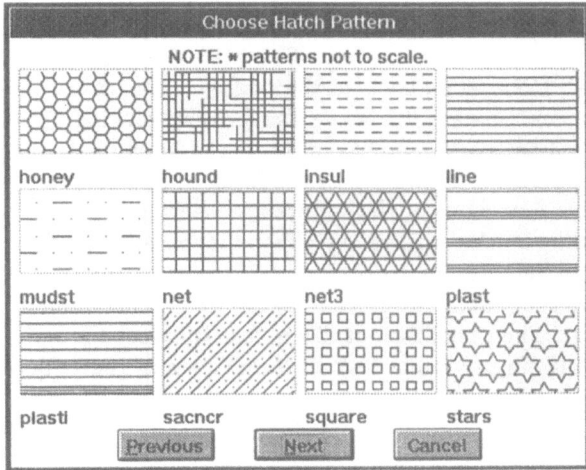

Figure 4.30 Selecting a hatch pattern

Move the cursor arrow into the rectangle of star shapes and press the mouse button. This selects the pattern and goes back to the Hatch Options dialogue box. The word "stars" now appears beside the Pattern button.

The next thing to do is check that the Hatching Style is set to "Normal". This means that if there are inner boundaries then the hatch will be applied to alternate areas starting with the outermost. We will make use of this to get the striped effect on the balloon. Of the other options for hatching style, "Outer" shades in only the outermost region. "Ignore" simply ignores any internal boundaries. To see how these affect the circle, square and triangle on the diallogue box you can pick the appropriate buttons. Make sure that you reset it to "Normal" before proceeding.

One last thing to do in the Hatch Options dialogue box is to select a scale for the pattern. As our drawing limits are relatively small, the default scale 1.00 will have to be reduced. Move the cursor into the Scale box and type **0.15**. Then click **OK**.

Back at the Boundary Hatch dialogue box pick **Select Objects**. Use a window to select the four lines drawn recently.

Select objects: **W**
First corner: **34,9**
Other corner: **@5,−5**
4 found. Select objects. **<ENTER>**

Once again the Boundary Hatch dialogue box appears. Before finally executing the command pick **Preview Hatch**. This will show how the shading will look. Press **<ENTER>** to go back to the dialogue box. If the pattern seemed satisfactory you can proceed to shade the box by picking **Apply**. Only on picking "Apply" is the command executed. If the hatching is not as shown in Figure 4.31, then ERASE it and try again. Make sure that you use the correct scale and that the four sides of the square are selected.

Now to shade in the stripes on the balloon. As we will be using the same pattern and settings as before we only need to select the object, preview and apply the hatch from the Boundary Hatch dialogue box. Pick **Draw/Hatch....** Then pick **Select Objects**. To select the stripes use a window and include the inner circle and all the ellipses of the balloon.

Command: _bhatch
Select objects: **W**
First corner: **10,20**
Other corner: **30,40,41**
4 found.
Select objects. **<ENTER>**
Press RETURN to continue

The shading shown in Figure 4.31 results because the ellipse boundaries have been selected within other boundaries. If you are happy with the hatching then press **<ENTER>** to go back to the dialogue box and pick **Apply** to execute the command.

Despite the hazard warning, there is little to fear from shading in objects with patterns. Take heed of the warning and follow the advice on saving the drawing. The reason for this caution is that if the pattern scale is small then it will use up a lot of memory. One single hatch command with an inappropriate scale can fill a 300 Mb disk! Remember ˆC will stop a rogue hatch in progress.

As with text, the more complicated the pattern the more it will slow down your redraw time. You will save time if you can leave any hatching as late as possible in the drawing. Freeze the layer when the hatching is not required to be displayed.

To finish this session, pick **File/Exit AutoCAD** followed by picking **Save Changes....**

Summary

You have encountered AutoCAD's most important drawing commands and entity types in this chapter. You have also used AutoCAD's entity selection procedure. This is common to many commands including ERASE and HATCH.

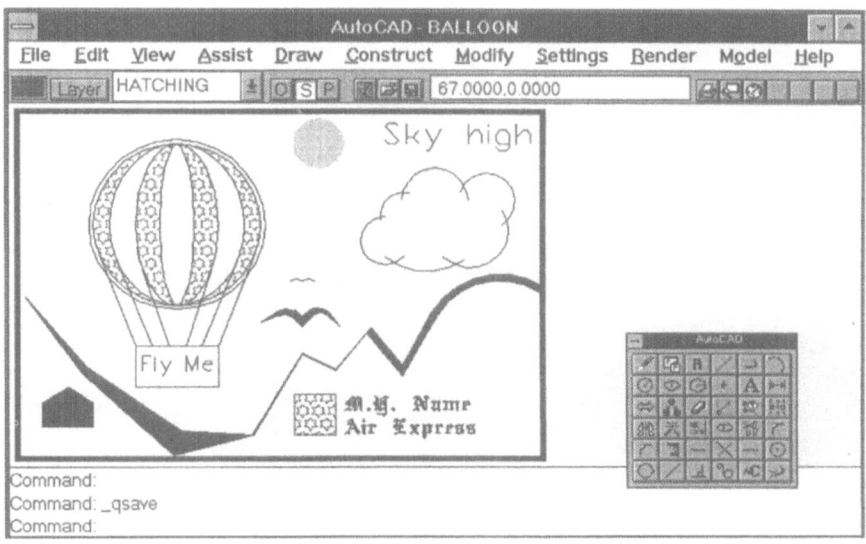

Figure 4.31 The AutoCAD Air Express

Polylines are versatile entities combining lines, arcs, traces and solids. You can speed up REDRAWs by turning FILL off, QTEXT on and freezing the hatch layers.

You should now be able to:

Select objects in many ways.
ERASE unwanted entities.
Restore items deleted in error.
Draw circles, arcs and ellipses.
Insert multiple lines of text.
Insert centred lines of text.
Define text styles using different fonts.
Draw rectangle.
Draw polylines with constant and varying widths.
Merge polylines with arcs.
Get Aerial Views
Create SOLID triangles, rectangles, circles and doughnuts.
Select a hatch pattern and perform a test shading.
Shade in multiple objects.
Speed up REDRAWs.

Chapter 5 CONSTRUCTIVE EDITING

General

The term constructive editing is used to describe those commands which either replicate existing entities or alter their characteristics. This chapter begins with some of the commands that alter the objects and then considers some of the ways to duplicate them. This will help to accelerate the repetitive type of work needed to complete the Eiffel tower (Figure 5.1) that was started in Chapter 3.

Start up AutoCAD, pick **File/New** and give EXPRESS2 as the default drawing name (Figure 5.2).

This copies all of the EXPRESS2.DWG file to EIFFEL.DWG and leaves the old file unchanged. Now set SNAP to 2.5, GRID to 5 and skip to the next section, "Drawing the Arch".

If you haven't got the EXPRESS2.DWG file or another file containing the drawing from Chapter 3, Table 5.1 summarizes what you will need.

Figure 5.1 Eiffel target drawing

Figure 5.2 Eiffel = EXPRESS2

Table 5.1 EXPRESS2.DWG Summary

Layer	Status	Color	Linetype
0	ON	WHITE	CONTINUOUS
CLINE	ON	WHITE	CENTER
CONST	FROZEN	RED	CONTINUOUS
EIFFEL	ON	WHITE	CONTINUOUS
MARGIN-TEXT	ON	WHITE	CONTINUOUS

Current layer: EIFFEL
SNAP = 2.5, GRID = 5.0
LIMITS from (0,0) to (65,45)

The center-line is drawn from (30,41.5) to (30,5) and is on the CLINE layer.
The following lines are all on layer EIFFEL.

Draw lines from (22.5,5) to (17.5,5) to (23.75,15) to (30,15)
From (23.75,15) to (23.75,17.5) to (30,17.5)
From (23.75,16) to (30,16)
From (25.3,17.5) to (26.5,22.5) to (30,22.5)
From (26.5,22.5) to (26.5,25) to (30,25)
From (26.5,23.5) to (30,23.5)
From (30,37) to (27.5,37) to (27.5,38) to (30,38)
From (27.5,37.5) to (30,37.5)
From (28,25) to (28,38) to (30,39.5) to (30,41.5)
From (29,37) to (29,38)

Drawing the arch

Before engaging in all this editing let's do a bit more preparatory work on the tower. Making sure that EIFFEL is the current layer, start by drawing the arch at the base of the tower. If you type the command you will have to input all the responses in bold but if you pick **Draw/Arc** followed by **Center, Start, Angle** from the pull-down menu you will only have to type the numbers. ARC can be found at the top of the DRAW sub-menu.

> Command: **ARC**
> Center/<Start point>: C Center: **30,5**   (A)
> Start point: **@7.5<90**   (B)
> Angle/Length of chord/<End point>: A Included angle: DRAG **90**

Note that while AutoCAD does remember the last command it doesn't necessarily remember all the options within the last command. So, to draw the second arc, pick **Draw/Arc/Center, Start, Angle** again.

> Command: **ARC**
> Center/<Start point>: C Center: **30,5**   (A)
> Start point: **@8.5<90**   (C)
> Angle/Length of chord/<End point>: A Included angle: DRAG **30**

Add some text but this time use the Dynamic TEXT command. Pick **Draw/ Text** and **Dynamic** from the pull-down menu or pick the capital "A" icon from the tool-box.

> Command: **DTEXT**
> Justify/Style <Start point>: **7,15**
> Height <3.00>: **1.8**
> Rotation angle <0>: **<ENTER>**

As you are prompted to input your text, a box should appear on the drawing at the insertion point. This box indicates the position of the next character and moves as you type the text. If you make a mistake while typing you can use the backspace key to remove the letters (as long as you haven't pressed <ENTER>).

> Text: **The Trifle <ENTER>**

It is clear from the drawing (Figure 5.3) that to include the full caption a second line of text is needed to avoid overwriting the actual tower. As you hit <ENTER> the DTEXT command automatically moves its box to the next line to allow more text to be input.

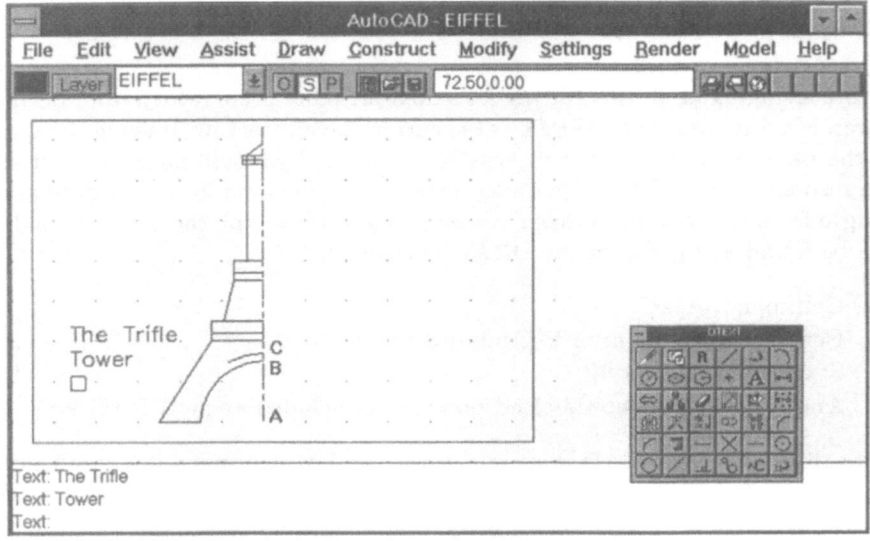

Figure 5.3 The Trifle Tower

Text: **Tower <ENTER>**
Text: **<ENTER>**

Finally to end the command press **<ENTER>** without typing any new text.
Don't worry that this text has not been put on the MARGIN-TEXT layer.
This sloppy practice will be corrected later.

A small corner of the drawing will be used as the prep area where the
metalwork for the tower will be assembled. **ZOOM** in and draw the structure
panels. The zoom command can be abbreviated to **Z**. Type Z followed by
<ENTER>. AutoCAD will echo with the full name and proceed as normal.

Command: **Z <ENTER>** ZOOM
All/Center/.../Window/<Scale(X/XP)>: **W**
First corner: **2.5,2.5**
Other corner: **9,6.5**
Command: **SNAP**
Snap spacing or ON/OFF/Aspect/Rotate/Style <2.50>: **0.25**

The basic structure panel in Figure 5.4 is a cross-braced frame. It is first drawn
to fit within a 1 by 1 box so that it can later be scaled to fit the different parts
of the tower.

```
Command: LINE
From point: 3,5                                                   (A)
To point: @1,1                                                   (B)
To point: @−1,0                                                  (C)
To point: @1,−1                                                  (D)
To point: <ENTER>
Command: <ENTER>
Line From point: 3.5,5                                           (E)
To point: @0,1                                                   (F)
To point: <ENTER>
```

Each landing will have a series of mini-arches comprising an arc and two lines each.

```
Command: <ENTER>
Line From point: 5,5                                             (G)
To point: @0,0.5                                                 (H)
To point: <ENTER>
```

Now type **ARC** and use the continuation property of this command.

```
Command: ARC
Center/<Start point>: <ENTER>
End point: DRAG @0.75<0                                          (J)
```

Pressing <ENTER> without giving a start point automatically selects the last line end point as the start of the arc and also makes the arc tangential to that line. A similar procedure can be done with the LINE command. Pick the LINE icon from the tool-box and press <ENTER> in response to the "From point:" prompt.

```
Command: LINE From point: <ENTER>
Length of line: 0.5                                              (K)
To point: <ENTER>
```

The "LINE continue" option automatically starts on the last line or arc end point. If an arc was drawn more recently than the last line then this new line will also be tangential to the arc. Hence, only the length of the new line needs to be specified.

The final component to be used is the fancy iron scrolling for the large arch at the base. For this you should use PLINE. The polyline icon is on the top row, second from the right in the tool-box.

```
Command: PLINE
From point: 7.5,5                                                (L)
```

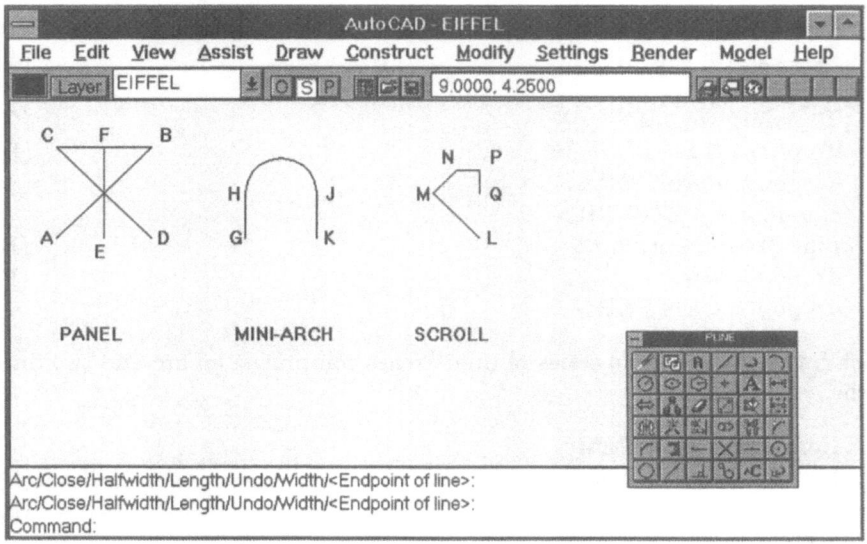

Figure 5.4 Structure panels

Current line-width is 0.0000
Arc/Close/. . ./Width/<Endpoint of line>: **@−0.5,0.5**   (M)
Arc/Close/. . ./Width/<Endpoint of line>: **@.25,.25**   (N)
Arc/Close/. . ./Width/<Endpoint of line>: **@.25,0**   (P)
Arc/Close/. . ./Width/<Endpoint of line>: **@0,−.25**   (Q)
Arc/Close/. . ./Width/<Endpoint of line>: **<ENTER>**

This should give you the three objects shown in Figure 5.4. As usual, the letters
on the diagram are only for reference and will not appear in your drawing.

Editing a polyline with PEDIT

Polylines are probably the most flexible entities in AutoCAD. This means
that they can also be cumbersome to edit. There are so many possibilities for
making changes that it is difficult to describe in a concise manner. In this
section you will encounter the more important facilities of Polyline EDITing
(PEDIT).

The PLINE scroll drawn above requires a few small changes. Firstly, it is
too small and secondly, it looks too square to be art nouveau. To make it a
bit bigger an extra point must be inserted and one vertex must be moved in
the polyline. Pick **Modify/Polyline Edit** from the menu bar.

Command: ai_peditm
Select objects: **7.5,5**   (L)
Select objects: **<ENTER>**
Close/Join/Width/Edit vertex/.../eXit <X>: **Edit**

Choose the edit vertex option. The prompt line changes and an X appears on
the polyline at its first point or vertex.

Next/Previous/.../Width/eXit <N>: **N**   (M)

The **N** selects the next vertex on the polyline and the "X" should now be at
point M (7.0,5.5). Now type **Insert** to put in a new point.

Next/.../Insert/.../Width/eXit <N>: **I**
Enter location of new vertex: **@.25<90**   (R)

The new shape should look like Figure 5.5(b) with the X still at point M. Move
the X to vertex, N (7.25,5.75), by typing **Next** twice. Once the X is at the
correct vertex the vertex can be moved to its new location (Figure 5.5c).

Next/Previous/.../Width/eXit <N>: **N**   (R)
Next/Previous/.../Width/eXit <N>: **N**   (N)
Next/.../Move/.../Width/eXit <N>: **M**
Enter new location: **@.25<90**   (S)
Next/Previous/.../Width/eXit <N>: **X**
Close/Join/.../Spline curve/.../Undo/eXit <X>: **S**
Close/Join/.../Spline curve/.../Undo/eXit <X>: **X**

Selecting the X option exits from the "Edit vertex" routine and returns you
to the PEDIT prompt, "Close/Join/...". To make the polyline into a smooth
curve select the Spline curve option by typing S. This executes a cubic B-
spline curve fitting routine. This type of curve gives an extremely smooth
shape fitted to the vertex points of the orginal polyline. The curve won't
actually pass through the vertices but will be drawn nearby. The technique is
named after the mathematician, Bezier, who invented it. Bezier is regarded by
many as the father of computer graphics and surface modelling. You can select
either a quadratic or cubic spline by setting the AutoCAD system variable
"SPLINETYPE" to 5 or 6 respectively. The quadratic version resembles the
original polyline shape more closely, while the cubic gives more "appealing"
curves.

Command: **SPLINETYPE**
New value for SPLINETYPE <5>: **6**

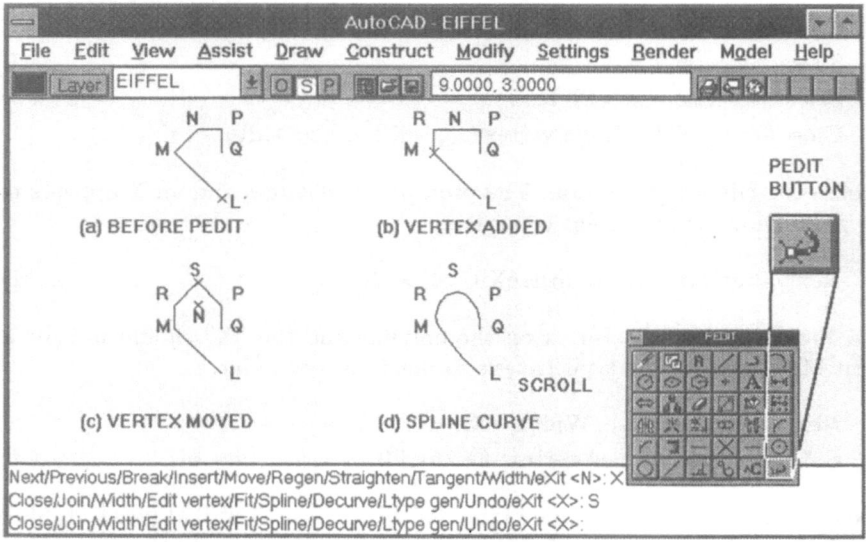

Figure 5.5 Polyline edit

Early versions of AutoCAD gave only the "Fit" option for smoothing polylines. This just changes all the line segments into arc segments. The arcs all pass through the original vertices. It is faster but visually inferior to the spline.

Other options for editing polylines include making it into a closed polyline, joining two polylines together and changing the width. You can also "Decurve" a spline or fitted curve.

The scroll work should now resemble the curlicue shown in Figure 5.5(d).

Moving objects

All AutoCAD entities can be modified to alter their position in the drawing. This gives great flexibility when making a drawing as you don't have to worry about getting everything to fit exactly. As the drawing progresses conflicts can be resolved by moving the objects to give a clearer picture.

In the above example you made use of ZOOM and "window" to draw the panels in close up. Now you can zoom out using AutoCAD for Windows' aerial view for a better view and move the panels into position, starting with the mini-arch. AutoCAD for DOS users should use ZOOM All followed by a suitable window. Click the **Aerial view** button indicated in Figure 5.6. The aerial view window then pops up. The little rectangle to the left of the tower

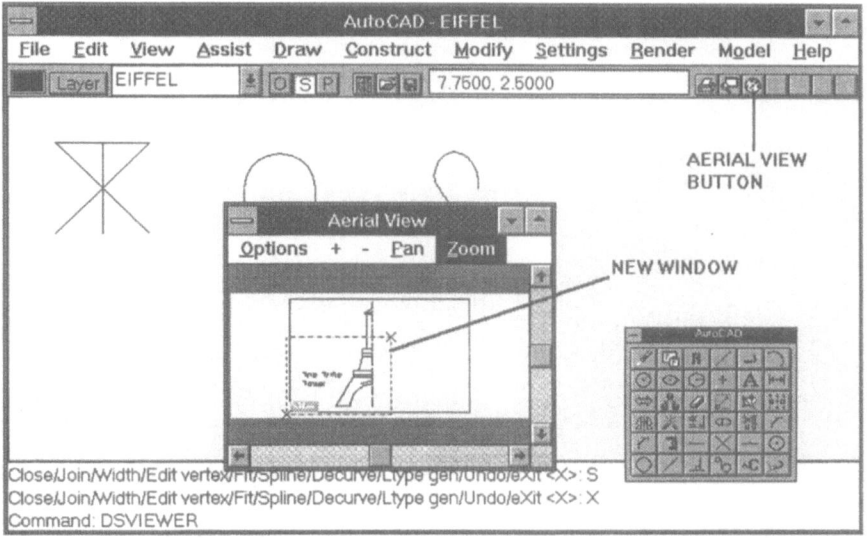

Figure 5.6 Aerial view of the tower

is the current drawing view. Move the cursor into the viewer window and pick
the two corners of the new zoom window.

The magnification will change in the background and the Aerial view
window remains active. If you are happy with the new magnification in the
drawing screen you can close the Aerial Window. To do this pick the **Control**
menu icon in the left corner just above "Options" and then pick **Close**.

The X-Y axes icon (UCS icon) in the lower left corner of the drawing
may now obscure the view of small panels. To remove the icon pick **Set-
tings/UCS/Icon**. This shows a short menu with a tick in front of the "On"
item. Pick **On** and both the tick and the UCS icon will disappear.

Pick **MOVE** icon from the tool box (Figure 5.7). The icon looks like a
double headed arrow pointing right and left. You will also find this command
in the Modify pull-down menu. AutoCAD enters the "Select objects:" mode
and you can make a window around the mini-arch. When all the selections
have been made you will be asked how far you want to move it.

Command: **MOVE**
Select objects: **Window**
First corner: **5,5**
Other corner: **@1,1**
3 found

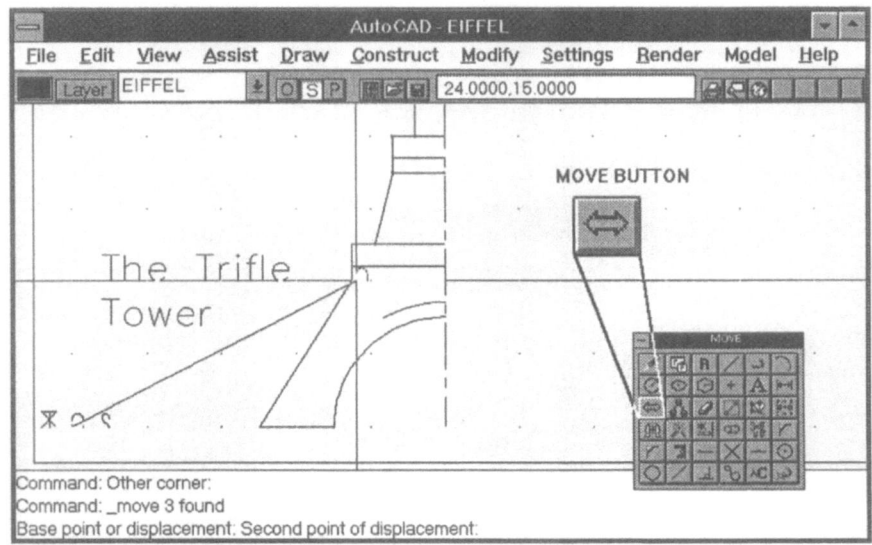

Figure 5.7 Moving the mini-arch

If the correct three objects, one arc and two lines, have been selected press **<ENTER>** to end the selection procedure.

Select objects: **<ENTER>**
Base point or displacement: **5,5**

This is the point from which the object is to be moved. You are then prompted for the new location. The mini-arch is then moved to the first landing (Figure 5.7).

Second point of displacement: **24,15**

The base point and second point do not have to be at the old and new locations. All that is important is the relative displacement between the two points. For example, the above movement would also have been achieved if the points (0,0) and (19,10) were input.

Base point or displacement: 0,0
Second point of displacement: 19,10

A third way of achieving the same result is to input the relative displacement at the first prompt and just press **<ENTER>** at the second.

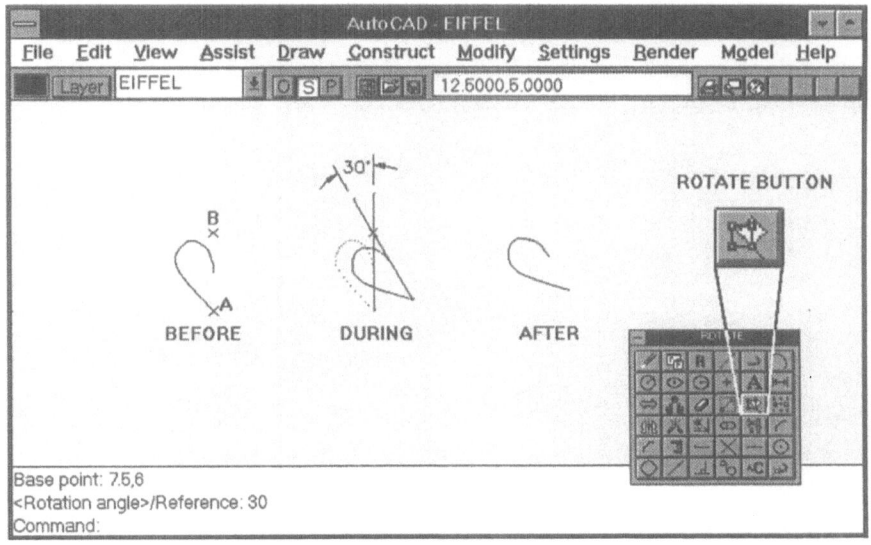

Figure 5.8 Rotating the polyline

Base point or displacement: 19,10
Second point of displacement: <ENTER>

Rotating objects

Before moving the scroll work to the main arch it must be put into the correct orientation. The scroll will be used to fill the area between the shorter and longer arcs. As the shorter one has an included angle of 30 degrees the scroll has to be rotated by that angle (Figure 5.8). To see the operation clearly, zoom in closely. As it is not so easy to pick small windows with the aerial viewer we will use the standard ZOOM command.

Command: **ZOOM**
All/Center/.../Window/<Scale(X/XP)>: **W**
First corner: **5,3**
Other corner: **12,8**

The ROTATE command is also on the Modify pull-down menu and also on the tool box. Click the **Rotate** button from the tool box. The object selection procedure is as before and when all the objects have been found you are asked for the center of rotation and the angle.

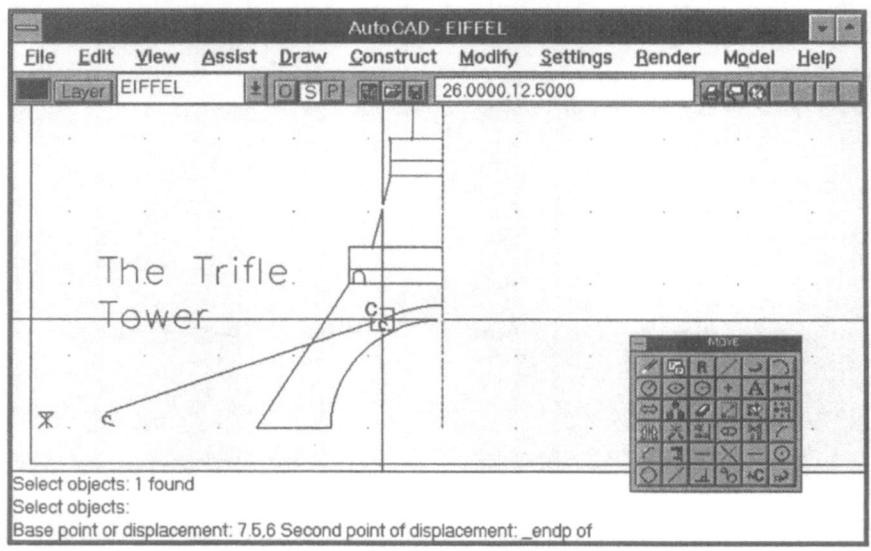

Figure 5.9 The scroll in the arch

Command: **ROTATE**
Select objects: **7.5,5**   (Pick the polyline at A.)
1 selected, 1 found.
Select objects: <**ENTER**>
Base point: **7.5,6**   (Center of rotation at B.)
<Rotation angle>/Reference: **30**

The scroll has been rotated 30 degrees anti-clockwise about the point (7.5,6).
Now do a "ZOOM Previous" and move the polyline to the end of the shorter
arc. The point (7.5,6) will be used as an imaginary handle by which the curve
will be moved to the point C shown in Figure 5.9.

Command: **Z** ZOOM
All/.../Previous/.../<Scale(X/XP)>: **P**
Command: **MOVE**   (Pick the MOVE button.)
Select objects: Pick scroll. 1 found.
Select objects: <**ENTER**>
Base point or displacement: **7.5,6**
Second point of displacement: **endp** of Pick upper arc at point C.

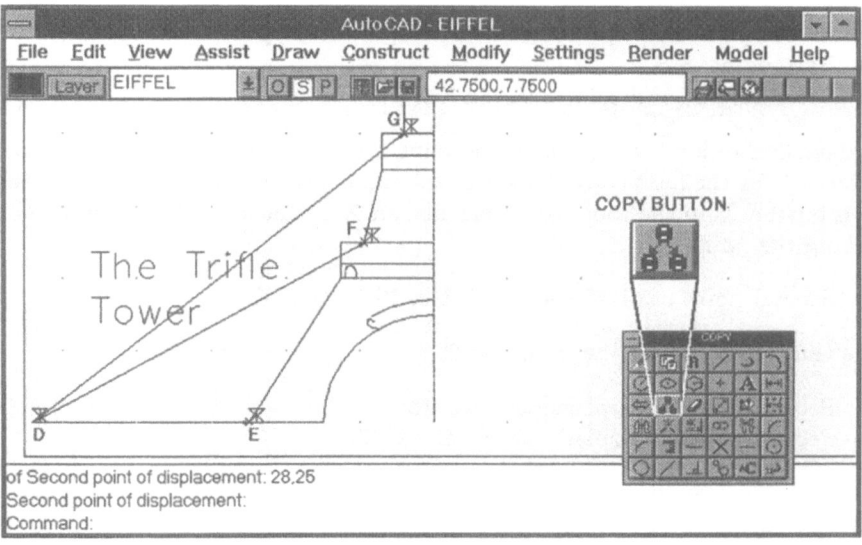

Figure 5.10 Multiple copies

Copying

The basic structure panel is still on the ground awaiting erection. Instead of simply moving it over you can copy it to each tier of the structure (Figure 5.10). The COPY command operates a bit like MOVE. You first select the objects to copy and then provide a displacement showing where you want them copied to. Unlike MOVE the original objects are left untouched and duplicates appear in the new positions.

In this particular operation you will make multiple copies of the panel, an extra feature of COPY. The first copy will be made to point E at the base, the second to point F and the third to the upper tier at G. The base point for all three copies will be the lower left corner of the panel, D (3,5).

> Command: **COPY**
> Select objects : **Window**
> First corner: **3,5**   (D)
> Other corner: **@1,1**
> 4 found

This surrounds the panel and finds its four lines.

> Select objects: **<ENTER>**

<Base point or displacement>/Multiple: **Multiple**
Base point: **3,5**   (D)
Second point of displacement: **17.5,5**   (E)

The panel should now appear at the point E and the prompt asks for "Second point..." for the next copy. This uses the same base point. Pick the **intersection** button from the tool box (looks like an X on the second last row) before picking the point F.

Second point of displacement: **int** of Pick point F.

The third copy is made to point G (28,25)

Second point of displacement: **28,25**   (G)
Second point of displacement: **<ENTER>**

When all the desired copies have been made press **<ENTER>** without giving any "Second point...".

The COPY command can also be used to make single copies. To do this give the coordinates of the "Base point or displacement" instead of typing "Multiple". It will then work as above but will ask you for a single "Second point..." before returning to the "Command:" prompt.

To draw the "inside leg" of the tower (Figure 5.11) you can copy the inclined lines. Copy the line EH to a point **3.3333** units to the right and the leg of the middle tier 2.5 to the right.

Command: **COPY**
Select objects: Pick line EH. 1 found.
Select objects: **<ENTER>**
<Base point or displacement>/Multiple: **3.3333,0**
Second point of displacement: **<ENTER>**
Command: **<ENTER>**
COPY
Select objects: Pick inclined line near F. 1 found.
Select objects: **<ENTER>**
<Base point or displacement>/Multiple: **2.5,0**
Second point of displacement: **<ENTER>**

Altering objects' characteristics

All the basic pieces of the tower are now in place. Something must be done to correct their sizes to make them all fit. The panels are too small and there is a

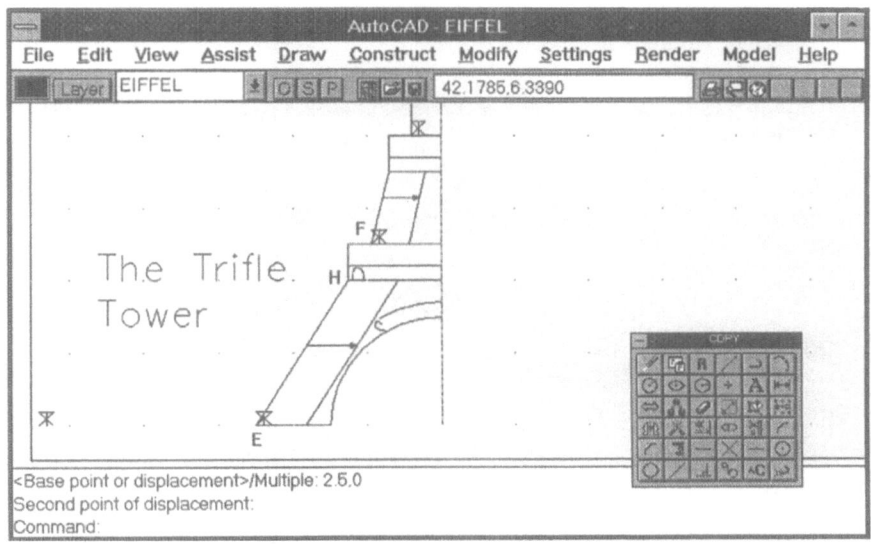

Figure 5.11 The inside leg

gap between the short arc at C and the inner part of the leg. Also the original panel is still at the point (3,5).

The first bit of tidying up is to sort out the use of layers. The text was put on the EIFFEL layer but ideally it should have been on MARGIN-TEXT. Also there was a small spelling error in the name of the tower! Can one command rescue the situation? Rather than delete the original panel it can be moved to the CONST layer.

The command to do this is called DDCHPROP or Change Properties. The "DD" indicates that this uses a dialogue box. Pick **Modify/Change/** followed by **Properties...** (Figure 5.12). You are then asked to select the objects. Use a window to collect the small panel.

Command: **DDCHPROP**
Select objects: **W**
First corner: **3,5**
Other corner: **@1,1**
4 found
Select objects: **<ENTER>**

At this point the Change Properties dialogue box appears (Figure 5.12). Click the **Layer** button and then pick the line:

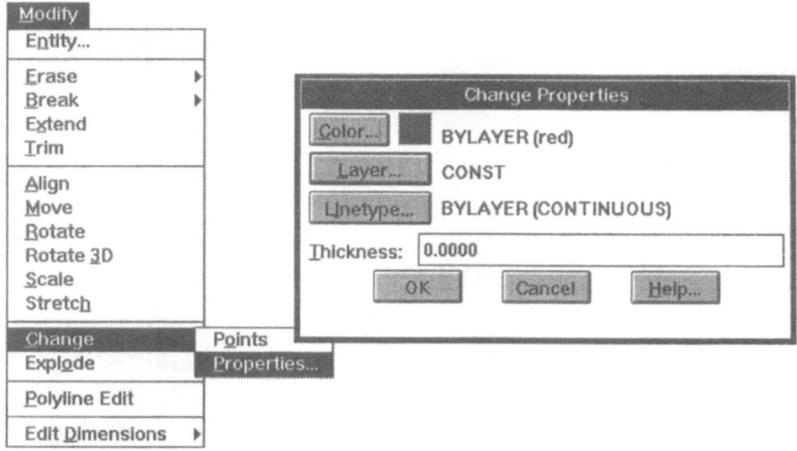

Figure 5.12 Changing of the layer

CONST On ... red CONTINUOUS

Then pick **OK** to set the new layer for the four entities. Note that the color and linetype will automatically be updated to the layer settings. Finally, click **OK** in the Change Properties Dialogue box.

There is also a non-dialogue box version of this command called, CHPROP, which users of older versions of AutoCAD will be familiar with. For example to do the previous operation you would type:

Command: **CHPROP**
Select objects: **Window**
First corner: **3,5**
Other corner: **@1,1** 4 found
Select objects: **<ENTER>**
Change what property Color/LAyer/LType/Thickness) **LAYER**
New layer <EIFFEL>: **CONST**
Change what property (Color/LAyer/LType/Thickness) **<ENTER>**

To correct the spelling error pick the "The Trifle" and to put it onto the MARGIN-TEXT layer use the more general Modify Entity procedure. The command DDMODIFY is executed by picking **Modify/Entity...** from the menu bar. You are prompted to select only one object to modify.

Command: _ddmodify
Initializing...DDMODIFY loaded.

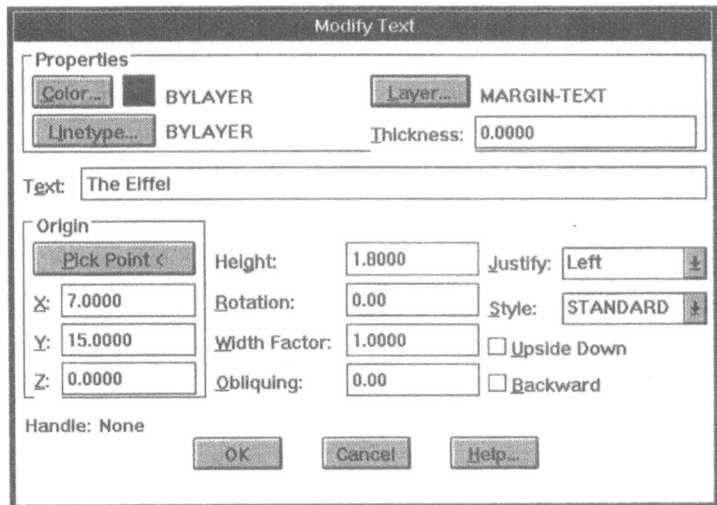

Figure 5.13 Modify Text dialogue box

Select object to modify: Pick the letter "f" in the word Trifle

This command then brings up a context-sensitive dialogue box. As the entity selected is a text entity the Modify Text dialogue box pops up (Figure 5.13). This contains all the parameters defining the text.

Click the **Layer** button in the dialogue box. Then pick the line:

MARGIN-TEXT On ... white CONTINUOUS

Then pick **OK** to get back to the Modify Text dialogue box. You can move to the line giving the value of the text, The Trifle, either using the mouse or by the hot-key. As the letter "e" is underscored in the word "Text" this is the hot-key. Typing **ALT+E** will automatically move the cursor into the text input box. The text will appear in reverse video with the cursor at the end. The cursor should appear as a blinking vertical line. Use the **left arrow** key to move along the text. When the blinking cursor is on the T of Trifle press the **Del** key on the keyboard six times to delete the word. Now type the word **Eiffel**. When you are happy with the modifications use the arrow shaped cursor to pick **OK**. The old text is then replaced by the new (Figure 5.14).

Using Modify/Entity... to correct typing errors in drawing text is much faster than erasing and re-typing. Above we needed to change two of the text entity's parameters, namely its layer and its value. We could also have altered

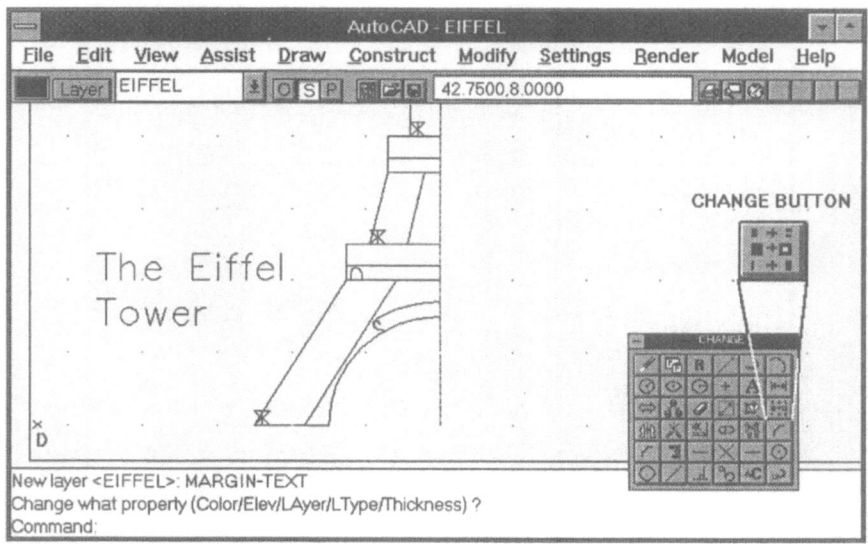

Figure 5.14 Changing text and layers

the size, justification, position, etc. If you just want to correct a spelling error there is another command called DDEDIT.

As this duplicates one of the facilities in DDMODIFY it has been omitted from the menu system. However, as most text modifications are necessitated by miskeying the AutoCAD Express recommends DDEDIT as a viable short cut. You must type the command, then select the text entity and edit it in the dialogue box in a similar manner to DDMODIFY.

Command: **DDEDIT**
<Select a TEXT or ATTDEF object>/Undo: Pick the text

You can move the cursor directly to the letter to change, replace it and click "OK" to execute the alteration. A benefit of DDEDIT over DDMODIFY is that DDEDIT ignores any non-text entities picked by mistake. On the other hand DDMODIFY can be used with any entity type and displays an appropriate dialogue box. With both commands you must select the entity by picking a point on it.

The Change point facility that appeared on the menu in Figure 5.12, allows the user to make clever geometrical edits to entities. It is of great use to those involved in digitizing drawings from paper originals. Inaccuracies in the digitizing process mean that some lines that should meet at a point either

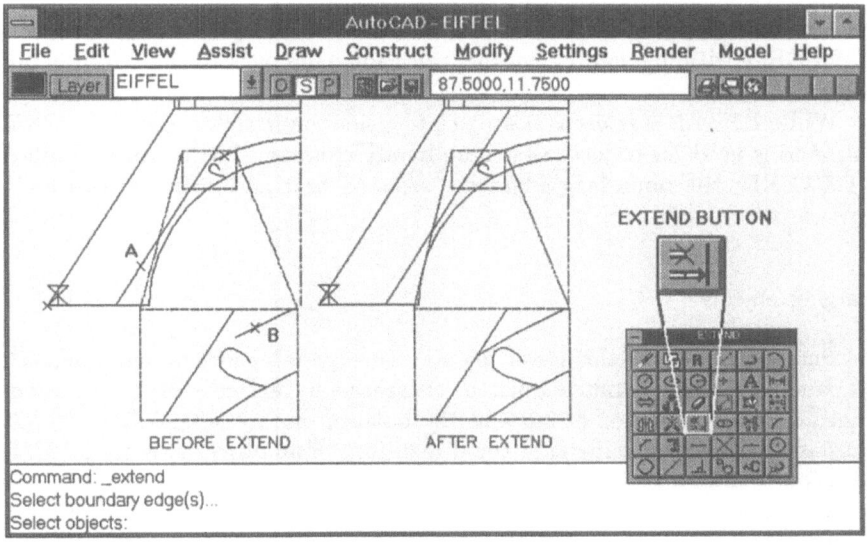

Figure 5.15 Extending the arch

don't meet or do so at the wrong point. The AutoCAD user can use CHANGE or its sub-set commands, EXTEND or TRIM, to force the lines to meet.

In EIFFEL there is a gap between the inner part of the tower's leg and the shorter of the two arcs in the arch. The EXTEND command can be used to change the arc's end point so that it meets the line exactly (Figure 5.15). Having picked **EXTEND** from the "Modify" menu you then must give Auto-CAD the boundary edges to which you want to extend the entity. In this case pick the inner line.

Command: **EXTEND**
Select boundary edge(s)...
Select objects: **22,7**   (A)
1 selected, 1 found

You will then be prompted for a second edge. In this example, only one edge is required. AutoCAD then asks for the objects to be extended. Pick the shorter arc and the curve is projected to meet the line.

Select objects: **<ENTER>**
<Select object to extend>/Undo: **26.5,12.75**   (B)
<Select object to extend>/Undo: **<ENTER>**

You can extend more than one item so AutoCAD prompts for more. Pressing <ENTER> without making another selection exits back to the "Command:" prompt.

While EXTEND is used to project to a new intersection point the TRIM command is used for objects that are already crossing. TRIM works similarly to EXTEND; the boundary edges are selected first and then the objects to trim.

Enlarging objects

In order to fit, each of the tower leg structure panels must be enlarged. The top panel, near the point G must be increased by a factor of 2, the middle panel near F by a factor of 2.5 and the bottom one by 3.3333. The SCALE command allows you to increase the dimensions of an object (Figure 5.16). Its operation is similar to that of ROTATE. You select the objects to SCALE and give a base point about which the objects will move when enlarging. Finally, you specify the magnification factor. The factor must be positive and not equal to zero. Giving a factor less than 1 reduces the size of the object while values greater than 1 increase it. The X and Y dimensions are changed by the same amount.

Using an initial panel size that fitted within a 1 by 1 box makes the calculation of the appropriate scale factors straightforward. To enlarge the panel at point E pick **SCALE** from the Modify menu. Then select the panel using **window**, give point G as the base point and finish up by giving the scale factor of **2**.

    Command: **SCALE**
    Select objects: **Window**
    First corner: **28,25**   (G)
    Other corner: **@1,1**
    4 found
    Select objects: <**ENTER**>
    Base point: **28,25**
    <Scale factor>/Reference: **2**

Now scale the panel at F. Pressing <ENTER> re-executes the last command.

    Command: <**ENTER**>
    SCALE
    Select objects: **Window**
    First corner: **25,17**   (near F)
    Other corner: **27,19**

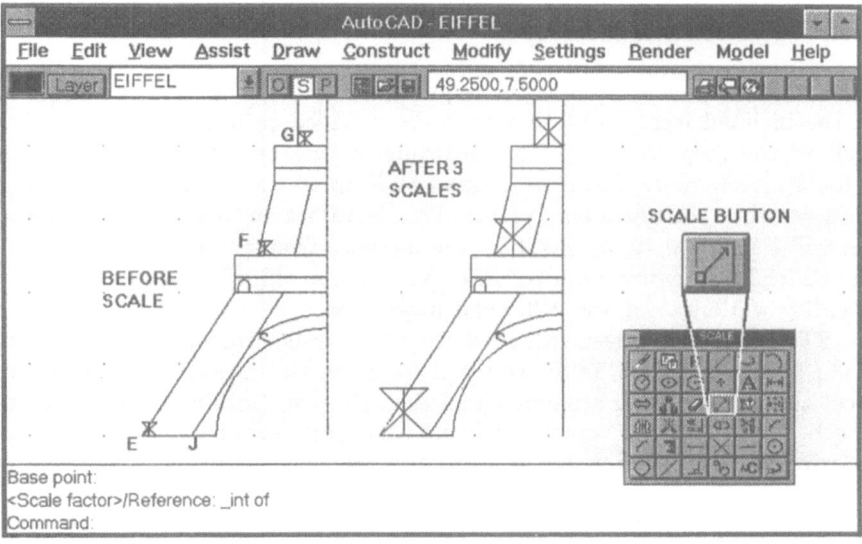

Figure 5.16 Scaling up the panels

4 found
Select objects: **<ENTER>**
Base point: **INTERSEC** of Pick point F.
<Scale factor>/Reference: **2.5**

And finally to scale the bottom panel. This time use the intersection points at
E and J to specify the scale factor. Command: **<ENTER>**

SCALE
Select objects: **Window**
First corner: **17.5,5**   (E)
Other corner: **@1,1**
4 found.
Select objects: **<ENTER>**
Base point: **17.5,5**   (E)
<Scale factor>/Reference: **R**
Reference length <1>: **<ENTER>**
New length: **INTERSEC** of Pick point J.

The reference length is the length of the original object. It can be picked by
giving two points on the object. In this case the default length of 1 is correct.

Stretching objects into shape

The two lower panels are now the required size but they are the wrong shape for the inclined legs. To fit they must be changed to become skew with the angle of the respective leg. Concentrating on the bottom panel first, zoom in for a closer view. To make it skew, the top of the panel must be moved sideways while the bottom stays put. That is, all the line end points in the top half will be shifted to the right by the distance from S1 to S2 (Figure 5.17). The STRETCH command, like EXTEND has the ability to act on one end of an entity while leaving the other end untouched.

STRETCH is easiest to operate if it is picked from the Modify menu. When you pick **STRETCH** you will be given the message that you must select the objects using a window and the "Crossing" option is automatically chosen. The "Crossing" option picks up all objects totally or partially within the window. This window has to contain all the end points to be shifted.

Command: **STRETCH**
Select objects to stretch by window or polygon ...
Select objects: **c**   (Crossing)
First corner: **17,7.5**   (W1)
Other corner: **21.5,9**   (W2)
5 found

Even the inclined leg becomes ghosted because it crossed through the window. As only the end points that were actually within the window are stretched, the leg will be okay. However, if you are in doubt as to whether an object might be undesirably stretched you can **Remove** it from the selection set. For practice remove the leg.

Select objects: **Remove**
Remove objects: **21.5,11.5**   (Point R1 on leg)
1 found, 1 removed
Remove objects: **<ENTER>**

When the selection/de-selection has been completed you are prompted for the point to stretch from and for the new destination.

Base point: **INTERSEC** of Pick point S1.
New point: **INTERSEC** of Pick point S2.

The panel should now fit snugly into the inclined leg (Figure 5.17).

Repeat this procedure to make the panel at point F fit its leg. Remember that only the end points that are within the window are stretched.

Command: **STRETCH**

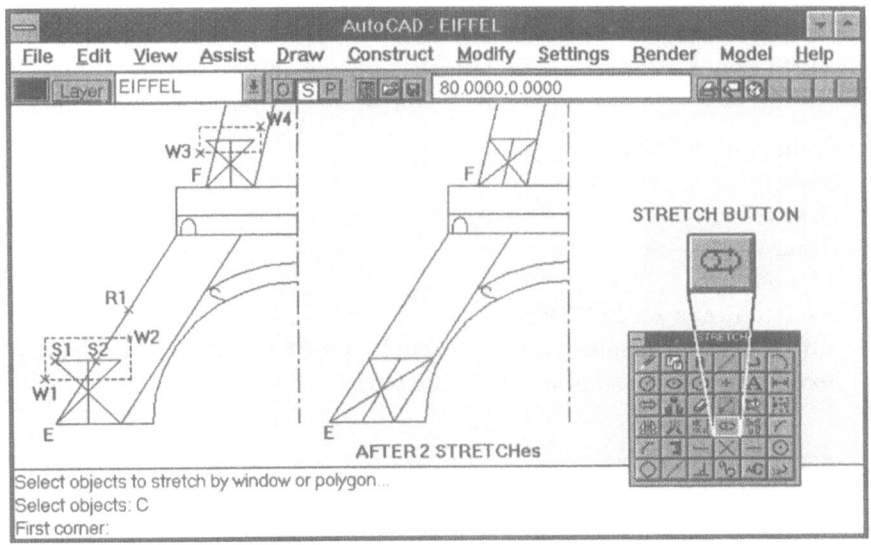

Figure 5.17 Stretching some points

Select objects to stretch by window or polygon ...

Select objects: **C**	(Crossing)
First corner: **25,19.5**	(W3)
Other corner: **28,20.5**	(W4)
5 found.	

If you don't use the "Crossing option" you will get the message "1 found" and you will have to add the unghosted lines of the panel.

Select objects: **<ENTER>**	
Base point: **INTERSEC** of **25,20**	(near the top left corner)
New point: **INTERSEC** of **26,20**	(near intersection with leg)

The COPY command has already been used to make multiple copies of the original structure panel. It will now be used to duplicate the panel on the middle tier near point F. The other tiers require multiple copies of their panels. As these copies are in regular patterns, AutoCAD's ARRAY command will be used. This allows objects to be copied in rows and columns and circular arrays. The scroll work will first be mirrored to complete the heart shape and copied in a circular array along the arch.

To copy the panel at F use **COPY Window** to select and use object snap **intersec** at points **F**, near (25.25,17.5), and near (26,20), to give the displacement.

Command: **COPY**
Select objects: **Window**
First corner: **25,17**
Other corner: **28.5,20.5**
4 found
Select objects: **<ENTER>**
<Base point or displacement>/Multiple: **INTERSEC** of Pick point F.
Second point of displacement: **INTERSEC** of Pick point (26,20).

Mirror image

Mirroring objects allows the user to take advantage of symmetries in the object being drawn. For example, only half of the tower is being drawn since it is symmetrical about the center-line. At the end of the chapter it will be mirrored to complete the picture.

To illustrate the MIRROR command the scroll in the arch will be reflected in a line at an angle of 120 degrees (remember that is was previously rotated by 30 degrees from the vertical and 30+90=120). Before mirroring zoom in for a better view. Pick **Mirror** from the **Construct** menu and select the scroll polyline. For the mirror line give the intersection point of the polyline and the lower arc as the first point and a relative displacement with an angle of 120 degrees for the second.

Command: **ZOOM**
All/Center/.../Window/<Scale(X/XP)>: **W**
First corner: **22.5,10**
Other corner: **29,14.5**
Command: **MIRROR**
Select objects: Pick any point on scroll.
1 found
Select objects: **<ENTER>**
First point on mirror line: **INTERSEC** of Pick point M1, Figure 5.18
Second point: **@10<120**

The length of the mirror line doesn't matter, just its orientation.

Delete old objects? <N> **<ENTER>**

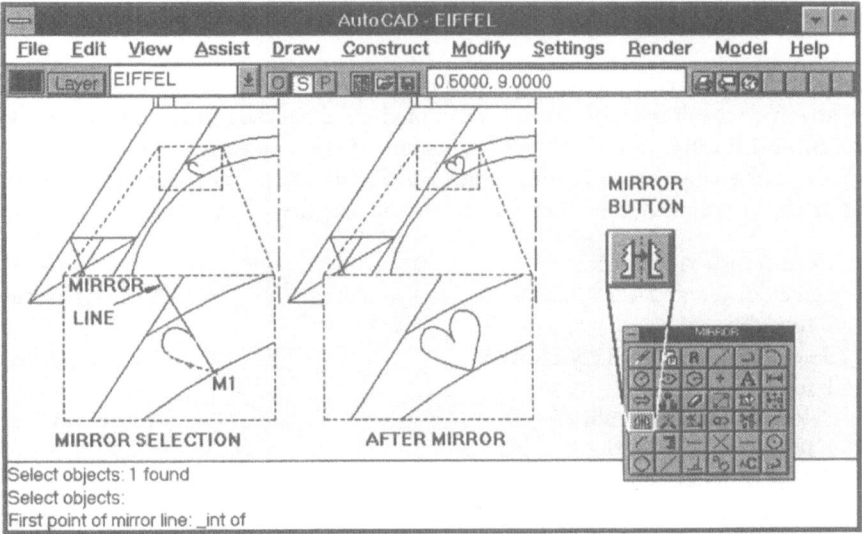

Figure 5.18 Taking polylines to heart

You do not wish to delete the original, so press **<ENTER>** and the heart is complete (Figure 5.18). At the end do a **ZOOM Previous** to get back to the last display magnification.

Command: **Z**

ZOOM

All/.../Previous/.../<Scale(X/XP)>: **P**

Multiple copies using ARRAY

In the tower drawing there are two simple patterns to be copied. Firstly, the mini-arches at the first landing will be copied to give two rows and six columns. Then the panels in the top section will be copied to give six rows and one column. Two slightly more difficult operations are involved to copy the hearts along the arch and the panels along the lower leg.

Rectangular arrays

With the ARRAY command you select the original objects to be arrayed. Then you specify whether they are to be copied in a rectangular grid or circular pattern and finally you give the dimensions of the pattern's repeated unit.

To make the rows of mini-arches in Figure 5.19, pick **Array** from the Construct menu and select the two lines and the arc of the original.

<pre>
Command: ARRAY
Select objects: 24.75,15.25   (A1 on right line)
1 found
Select objects: 24.375,15.875   (A2 on arc)
1 found
Select objects: 24,15.25   (A3 on left line)
1 found
Select objects: <ENTER>
Rectangular or Polar array (R/P)<R>: R
</pre>

This selects the rectangular grid pattern. You are now prompted for the number of rows and columns.

<pre>
Number of rows (---) <1>: 2
Number of columns (| | |) <1>: 6
</pre>

The "(---)" is to remind you that the rows are always horizontal and the "(| | |)" is for the vertical columns. Now input the distance between the rows and the columns.

<pre>
Unit cell or distance between rows (---): 1
Distance between columns (| | |): 1
</pre>

The mini-arch should be repeated to give a total of 12 arches. Note that the distance between the rows and columns is the length between a point on the original object and the corresponding point on its immediate neighbours. Inputting positive distances causes the duplicates to appear to the right and above the original. To make them appear on the left give a negative distance between rows. Similarly a negative distance between the columns causes the new objects to be drawn below the original.

To ARRAY the panel in the top part of the structure try the following:

<pre>
Command: ARRAY
Select objects: Window
First corner: 28,25   (G)
Other corner: @2,2
4 found
</pre>

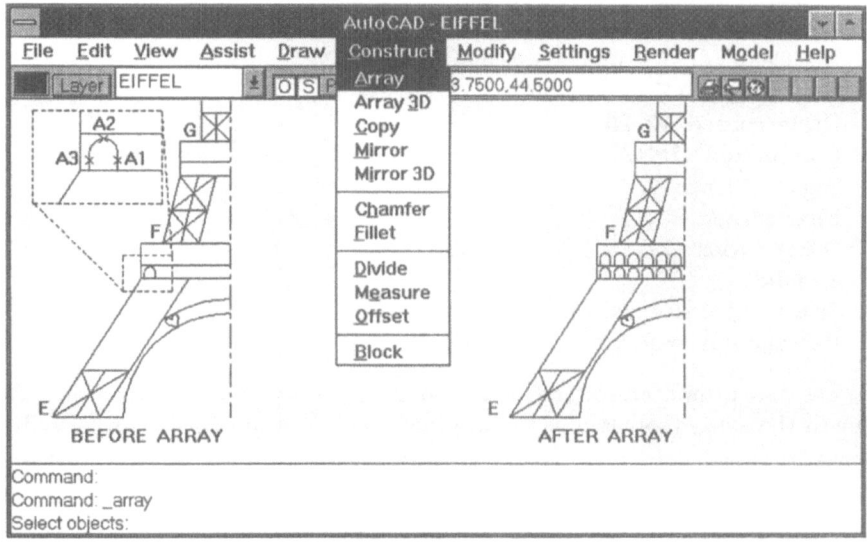

Figure 5.19 Rectangular array

Select objects: **<ENTER>**
Rectangular or Polar array (R/P)<R>: **<ENTER>**
Number of rows (---) <1>: **6**
Number of columns (| | |) <1>: **<ENTER>**
Unit cell or distance between rows: **2**

This should fill up the top of the structure with a total of 5 copies plus the original panel.

Circular arrays

The circular or polar array option is used to copy objects around some central focus point. For example, the spokes on a bicycle wheel could be drawn by copying one line in a circular pattern centerd on one of the end points. In the EIFFEL tower drawing the heart shapes will be copied along the arch. This can be done by using a polar array centerd on the point O (at the arc center) and repeating the object through 30 degrees, Figure 5.20.

Zoom in for more detail. Pick **ARRAY**, select the two halves of the heart shaped scroll and then opt for the **Polar** array.

Command: **Z**

```
ZOOM
All/Center/.../Window/<Scale(X/XP)>: W
Firsr corner: 17,4
Other corner: 32,15
Command: ARRAY
Select objects: W
First corner: 25,11                                          (W1)
Other corner: 27,13                                          (W2)
2 found
Select objects: <ENTER>
Rectangular or Polar array (R/P): P
```

You are now prompted for the center point of the array, the number of items to be in the array (copies plus the original) and the number of degrees to fill.

```
Center point of array: 30,5                                   (A)
Number of items: 5
Angle to fill (+=ccw, −=cw) <360>: −30
Rotate objects as they are copied? <Y> <ENTER>
```

The minus sign indicates a clockwise angle (as indicated by the "−=cw" in the parentheses). A positive angle would cause the copies to appear anti- clockwise of the orginal. The final prompt asks you if the objects are to be copied in their current orientation or if they are to be rotated. In this case they should be rotated.

Non-orthogonal rectangular arrays

In rectangular arrays above, the rows were vertically above each other. Similarly the columns are separated by horizontal distances. This is because the ARRAY directions are always parallel and perpendicular to the SNAP angle. The default snap angle is zero. Using **SNAP** you can rotate the snap angle to 58 degrees (the angle of the lower leg). Then the ARRAY command can be used to copy the panel along the leg.

```
Command: SNAP
Snap spacing or ON/OFF/Aspect/Rotate/Style <0.5000>: Rotate
Base point <0.0000,0.0000> 17.5,5                             (E)
Rotation angle <0.00>: 58
```

As you select a new snap angle the cursor cross-hairs and grid rotate by 58 degrees (Figure 5.21). Note that the drawing coordinates have not changed. It's just the snap locations that have altered. You can use ARRAY to copy along the 58 degree leg. This array has one row and three columns as the leg

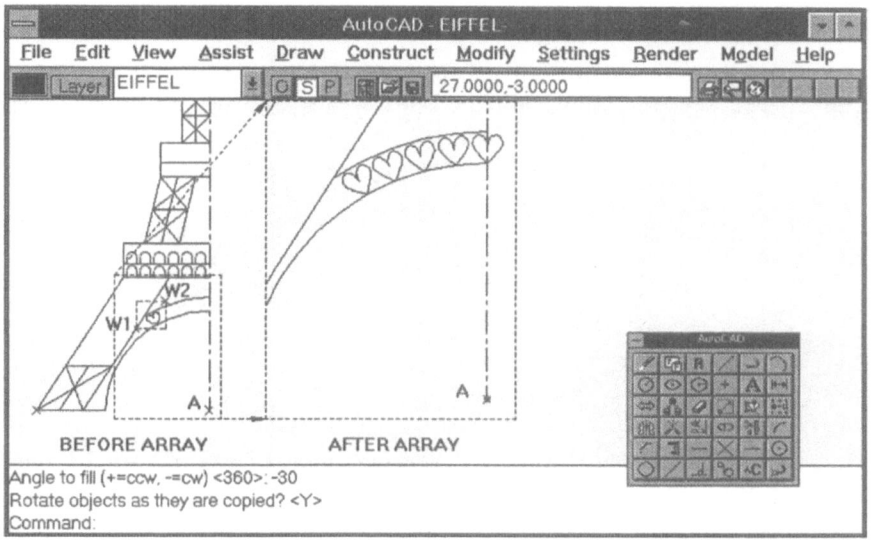

Figure 5.20 Polar array

direction is the effective horizontal. The distance between the columns can be found by object snapping to points E and N.

Command: **ARRAY**
Select objects: **Window**
First corner: **17.5,5** (E)
Other corner: **23,9**
5 found.

The horizontal line at the bottom of the leg may be selected. If so remove it.

Select objects: **Remove**
Remove objects: **20,5** (point on horizontal line)
Remove objects: **<ENTER>**
Rectangular or Polar array (R/P)<P>: **R**
Number of rows (---) <1>: **1**
Number of columns (| | |) <1>: **3**
Distance between columns: **INTERSEC** of Pick point E.
Second point: **INTERSEC** of Pick point N.

Now reset the SNAP angle back to zero.

Command: **SNAP**

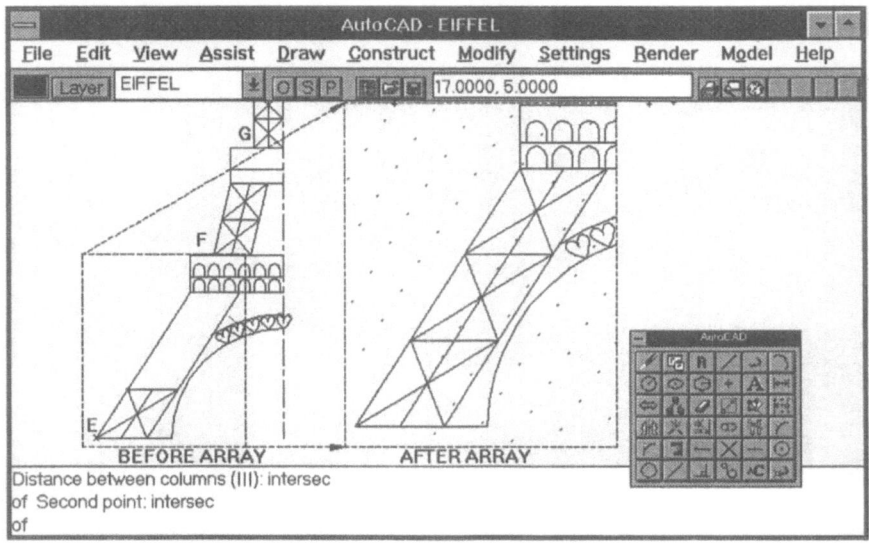

Figure 5.21 Rotated rectangular array

Snap spacing or ON/OFF/Aspect/Rotate/Style <0.5000>: **Rotate**
Base point <17.5000,5.0000> <**ENTER**>
Rotation angle <58.00>: **0**
Command: **Z**
ZOOM
All/Center/.../Window/<Scale(X/XP)>: **A**

Finishing up

The tower is now near completion. All that remains is to copy the mini- arches
to the second level and to mirror everything about the center-line. As this
particular operation involves a lot of entities it will be fairly computationally
intensive.

Command: **COPY**
Select objects: **Window**
First corner: **24,15**
Other corner: **26.75,17**
18 found
Select objects: <**ENTER**>

<Base point or displacement>/Multiple: **24,15**
Second point of displacement: **27,22.5**

Before executing any big copying operation you should save the drawing. Pick
File/Save... from the pull down menu. This executes a "quick save". This
saves to the current file name without requesting a confirmation.

Command: _qsave

Now you can mirror the left half of the tower including the text.

Command: **MIRROR**
Select objects: **Window**
First corner: **5,5**
Other corner: **30,40**
131 found

Only the center-line and the aerial on the roof should remain un-ghosted. If
the actual number of objects found on your drawing is much greater than 131
it means that you probably have copied some items onto themselves.

Select objects: <**ENTER**>
First point on mirror line: **30,5**   (A)
Second point: **@35<90**
Delete old objects? <N> <**ENTER**>

Again, the length of the mirror line doesn't matter, just its location and direction.
    Did the text become inverted? AutoCAD supports two modes of mirroring
text. The normal mode is for it to be inverted like everything else. Many times
this can yield silly results. The second mode which prevents all text from
becoming reversed is invoked by setting an AutoCAD system variable. The
variable is called MIRRTEXT and when its value is set to zero the mirrored
text will not be inverted.

Command: **MIRRTEXT**
New value for MIRRTEXT <1>: **0**

Now ERASE the inverted text and repeat the MIRROR command for the
original text only.

Command: **ERASE**
Select objects: **45,15**   (Approx. top line of text)
1 found
Select objects: **51,12**   (bottom line of text)
1 found
Select objects: <**ENTER**>

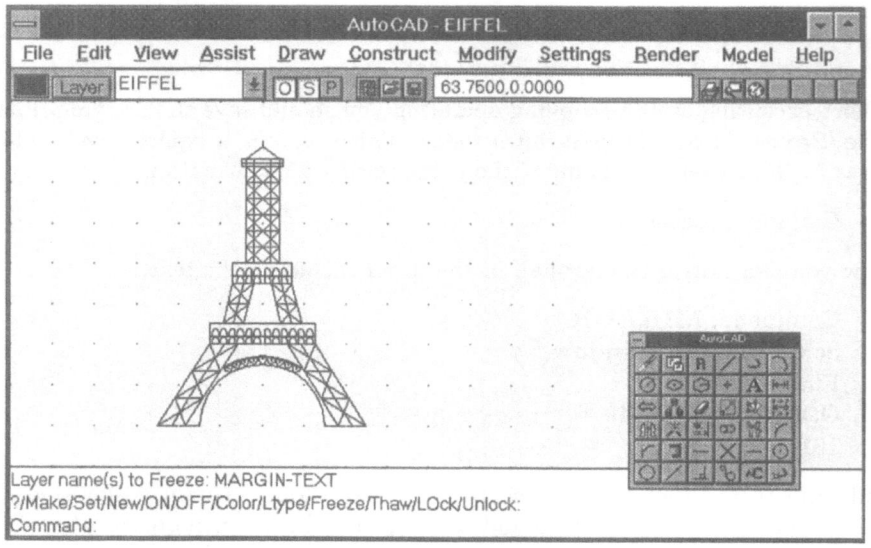

Figure 5.22 The completed Eiffel tower

Command: **MIRROR**
Select objects: **14,15**   (top line of text)
1 found
Select objects: **9,12**   (bottom line of text)
1 found
Select objects: <**ENTER**>
First point on mirror line: **30,5**
Second point: **@1<90**
Delete old objects? <N> <**ENTER**>

Finally, freeze all the layers excepts EIFFEL.

Command: **LAYER**
?/Make/.../Freeze/Thaw: **Freeze**
Layer name(s) to Freeze: *****

The asterisk is AutoCAD's wildcard and means all the layers in this case. However, the current layer cannot be frozen and AutoCAD gives the message:

Cannot freeze layer EIFFEL. It is the CURRENT layer.
?/Make/.../Freeze/Thaw: <**ENTER**>

Everything but the actual tower should become invisible as shown in Figure 5.22. Note that the two sets of text also disappear. This is because the MIRROR command preserves the layer of the entity. So even though you were working with EIFFEL as the current layer you could copy things on other layers. Furthermore the copied items are always put on the same layer as their original.

To finish the session, save the drawing and to exit AutoCAD type **END** or pick **File/Exit AutoCAD** from the pull-down menu.

Command: **END**

Summary

In this chapter you have been introduced to most of AutoCAD's editing commands. Some new aspects of other commands have also been covered.

You should now be able to:

Edit polylines and create smooth curves.
Move, rotate and copy objects within a layer.
Make multiple copies of objects in regular patterns.
Alter text.
Move things from one layer to another.
Change the proportions of objects.
Use the MIRROR command.
Set the MIRRTEXT variable.

Chapter 6 SUPER-ENTITIES

General

A number of entities can be grouped together to form a single new super-entity. In a way polylines are super-entities since they cause a number of lines and arcs to be grouped together. AutoCAD provides a more general way of linking objects together to form *blocks*. Blocks are mini-drawings that can be called up or inserted into a drawing at any location and as many times as desired. By compounding entities to form frequently used shapes or symbols the AutoCAD user can avoid unnecessary duplication of drafting. Blocks can be made globally available to drawings allowing AutoCAD users to build up libraries of complicated shapes which can be easily incorporated into any new drawing. Indeed, the true benefits of AutoCAD only become apparent when you have such libraries set up. Large assembly drawing can be quickly created by inserting the standard details from your library.

Blocks are more than just a stored shape. They can also contain non-drawing information such as a part number or a cost for the item. This extra information can be accessed to provide a bill of materials which could significantly improve the accuracy and speed of your estimates for the job.

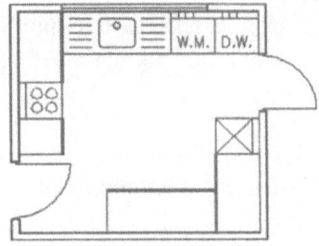

Figure 6.1 Kitchen fitted with blocks

Table 6.1 Layer settings for drawing KITCHEN

Layer name	State	Color	Linetype
0	On	7 (white)	CONTINUOUS
CONST	On	2 (yellow)	CONTINUOUS
FITTINGS	On	7 (white)	CONTINUOUS

Current layer: 0

Making a block

The exercise in this chapter uses the fitted kitchen industry to illustrate how to create and use the blocks in Figure 6.1. This will also cover adding text information to blocks and extracting this information. While making up some suitable objects for AutoCAD Express Kitchens Ltd you will encounter a couple of new editing commands, FILLET and OFFSET. The chapter finishes off with a look at some of AutoCAD's inquiry commands which give information about the drawing and also about the computer.

To start with you should begin a new drawing, calling it KITCHEN. The operating unit for the drawings in this chapter is the millimetre and so the limits must be set for an upper right corner of (6500,4500). The layers should be set up as given in Table 6.1. When this is done and the snap, grid and axis are set you can begin on the first block. This is to be the symbol for a door and comprises some lines and an arc.

>   Command: **LIMITS**
>   Reset Model space limits:
>   ON/OFF/<Lower left corner> <0.0000,0.0000>: **<ENTER>**
>   Upper right corner <420.0000,297.0000>: **6500,4500**
>   Command: **<ENTER>**
>   LIMITS
>   Reset Model space limits:
>   ON/OFF/<Lower left corner> <0.0000,0.0000>: **ON**
>   Command: **Z**
>   ZOOM
>   All/Center/.../Window/<Scale(X/XP)>: **A**

Now set the snap value to 100 and the grid at 200.

>   Command: **SNAP**

Snap spacing or ON/OFF/Aspect/Rotate/Style <1.0000>: **100**
Command: **GRID**
Grid spacing(X) or ON/OFF/Snap/Aspect <0.0000>: **2X**

A standard size door in a dwelling house has an opening approximately 800mm wide. Allowing for the frame, the door itself will make an arc with a radius of 750mm.

Command: **LINE**
From point: **1000,1000**   (A)
To point: **@25,0**   (B)
To point: **@0,750**   (C)
To point: **<ENTER>**
Command: **ARC**
Center/<Start point>: **@**   (This selects the last point, C.)
Center/End/<Second point>: **C**   (The center point, B.)
Center: **1025,1000**   (The intersection of the two lines.)
Angle/Length of chord/<End point>: **Angle**
Included angle: **-90**

The negative angle is required since the positive direction for angles is anti-clockwise.

Command: **LINE**
From point: **1800,1000**   (D)
To point: **@-25,0**   (E)
To point: **<ENTER>**

This is a door which is hinged on its left-hand side. The arc indicates the area swept by the door as it is opened. The assembly is now ready for BLOCK-ing (Figure 6.2). Pick **Block** from the **Construct** menu. You will then be prompted to name the block, give an insertion base point and after that you select the objects to be included in the block. The insertion base point is the reference point on the object by which it will be located. It is in effect the origin point for the block.

Command: **BLOCK**
Block name (or ?): **DOOR**

The rules for naming blocks are the same as for naming LAYERS: up to 31 characters long containing letters, numbers and the characters "$", "-" and "_". No spaces are allowed in the block name.

Insertion base point: **1000,1000**   (The left-hand end of the door, A.)
Select objects: **Window**

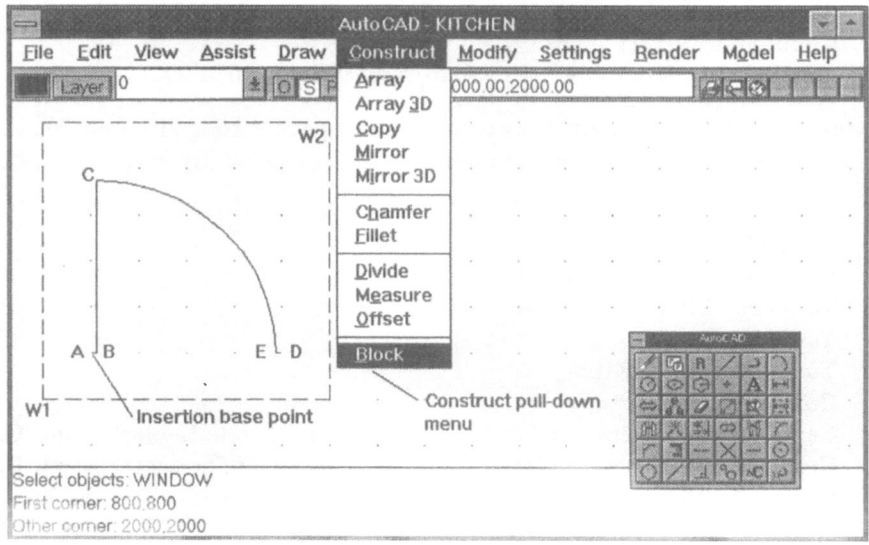

Figure 6.2 Blocking the door

First corner: **800,800**   (W1)
Other corner: **2000,2000**   (W2)
4 found

The four objects, three lines and an arc should become ghosted indicating that
they have been selected. End the selection procedure.

Select objects: **<ENTER>**

All four objects should now disappear from the screen, leaving only the blips
marking the window corners. To check that it is still part of the drawing,
execute the BLOCK command once more and reply to the block name prompt
with a "?". This will give a list of all the blocks in the current drawing. If you
wish to restore the objects that have disappeared use the OOPS command.

Command: **<ENTER>**
BLOCK Block name (or ?): **?**
Block(s) to list **<*>**: **<ENTER>**

The screen flips to text mode giving the information in Figure 6.3. Any block
that is listed can be INSERTed in the drawing as described in the next section.
To get back to the graphics screen you can close the "AutoCAD Text" win-

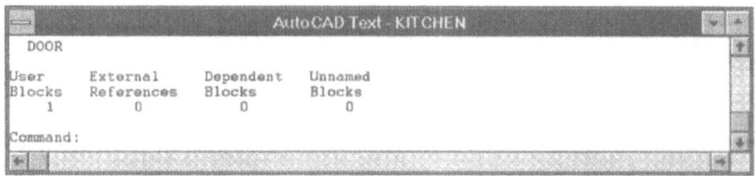

Figure 6.3 List of blocks

dow in the usual Windows fashion. Alternatively, you can execute the Redraw command by typing "R" followed by pressing <ENTER>. AutoCAD for DOS users can use the F1 key. However, in Windows the F1 key is for HELP.

Command: **R**
REDRAW

Inserting blocks

To draw the block pick Draw/Insert from the menu bar. This is almost the reverse of the blocking procedure. A dialogue box pops up (Figure 6.4) and you are asked for a block name to insert and where to place its insertion base point. There is considerable flexibility with inserting blocks. You can position it anywhere. You can alter its scale in either the X or Y direction and also change its orientation.

For the first insertion, move the cursor to the block name input field, click the mouse button and type **DOOR**. Make sure that the X appears opposite "Specify Parameters on Screen". All the other items are ghosted indicating that they cannot be set right now. Now pick **OK**.

If the DRAGMODE is on you should now be able to drag the image of the door around the drawing. If the door doesn't appear as you move the cursor then type **DRAG** followed by pressing <ENTER> in response to the insertion point prompt. You can then position the door either by picking a point or by typing the coordinates.

Command: ddinsert
Insertion point: **DRAG <ENTER> 2000,2000**   (Insertion A)

As soon as you input the insertion point the command prompts you for a scale factor to be applied in the X direction. The door may seem to disappear at this stage or it may be magnified on the screen. This is because AutoCAD is using the position of the cursor to calculate the scale factor. If the cursor is

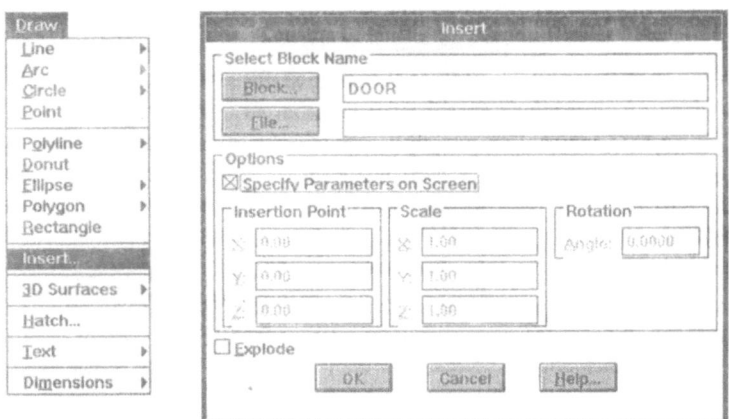

Figure 6.4 Insert menu and dialogue box

moved from (2000,2000) to (2100,2000) the scale factor is being taken as 100. In most cases it is safest to input the scale factor by keying in the number. The default scale factor is 1 which will draw the block at the same size as when it was defined. Press **<ENTER>** to accept the default. A different Y scale can be used to elongate or squash the object. The default situation is to use the same scale factor for both X and Y.

X scale factor <1>/Corner/XYZ: **<ENTER>**
Y scale factor (default=X): **<ENTER>**

Finally, you are asked for a rotation angle. A non-zero angle will cause the door to be rotated about its insertion point. You can also use the cursor to drag the door into the desired orientation. For this example press **<ENTER>** for zero rotation.

Rotation angle <0>: **<ENTER>**

The door should now appear in a position similar to "Xscale 1, Yscale=1, Angle=0" in Figure 6.5. If it didn't, try the command again but this time type all the responses fully. If you use the keyboard arrows to position the cursor then pressing **<ENTER>** picks the cursor location and not the default value.

To insert a door at B in Figure 6.5 with the hinge on the right-hand side use an X scale of −1 and Y scale of +1. This time specify all the values in the dialogue box (Figure 6.6). Pick **Draw/Insert**. The last block to have been inserted should already appear in the dialogue box, ie **DOOR**. Now click the "X" beside "Specify Parameters on Screen". The insertion point, scale and rotation fields should now appear normal. Move the cursor to the Insertion

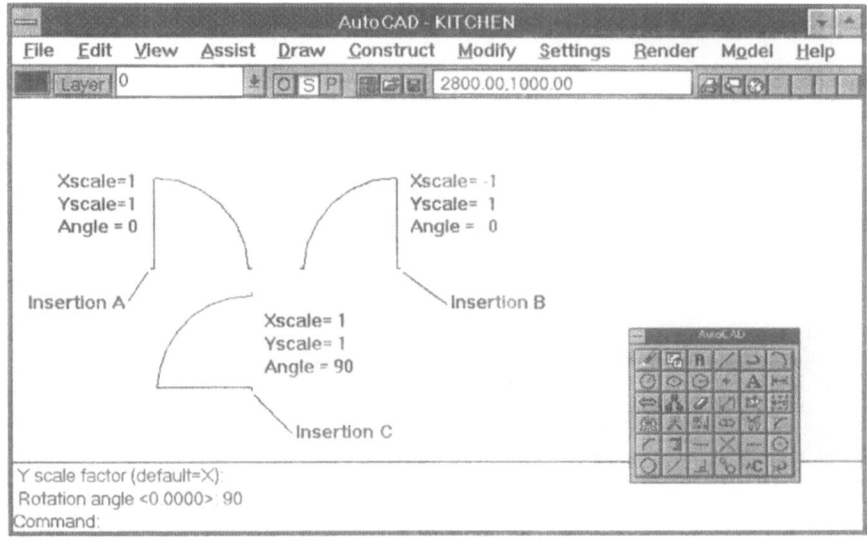

Figure 6.5 Inserting doors

Point, X field, click the mouse button and type **4000**. Similarly, put **2000** in the Y field. Then move to the X Scale box and type **−1**. The Y and Z scales may be automatically set to −1 so you have to reset them to +1 before picking **OK**.

Finally, to insert the door at C in Figure 6.5 at right angles to the other two use a rotation angle of 90 degrees (positive = anti- clockwise). This time, rather than use the dialogue box try inputting the information with the keyboard. At the command prompt type **INSERT** as follows:

Command: **INSERT**

Block name (or ?) <DOOR>: <**ENTER**>

Insertion point: **2800,1000**

X scale factor <1>/Corner/XYZ: <**ENTER**>

Y scale factor (default=X): <**ENTER**>

Rotation angle <0>: **90**

In many cases it is quicker to use the keyboard for block insertion rather than the dialogue box. You will need to experiment with the different methods to see which one you prefer.

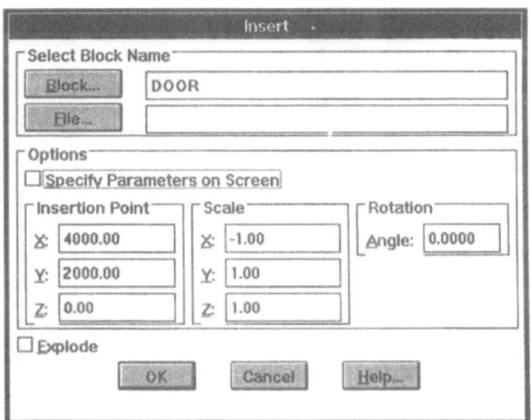

Figure 6.6 Inserting door at B

Global blocks

As mentioned above, blocks can be considered as mini-drawings. The converse is also true. Drawings themselves can be considered as large blocks. Indeed a whole drawing can be inserted into the current drawing using the Insert dialogue box in the normal way. Instead of giving a block name you give the drawing file name (and the DOS path if it is in another DOS directory but without the ".DWG" extension). The file name input field is just below the block input field. The insertion base point will normally be the origin but can be set to any point. Thus, all drawings are also blocks. As they are available to be inserted in any other drawing I call this type of block the "global block".

Blocks created in the same way as the DOOR above are only available within the drawing in which they were defined. They can be converted into global blocks, available to all drawings by using the **WBLOCK** command. This makes a copy of the block to a standard AutoCAD drawing file. WBLOCKs retain their layer, color and linetype settings. To write the DOOR block to a drawing file called "WDOOR.DWG" use the following command sequence.

The Create Drawing File dialogue box will appear. Go to the "File name:" box, click the cursor and type **WDOOR**. Then click "OK" and supply the block name.

Command: **WBLOCK**
File name: **WDOOR** or use the dialogue box
Block name: **DOOR**

The block is copied to WDOOR.DWG. The original is still intact in the current drawing. Only layers that are actually used in the block are retained in WDOOR. In this case all the entities were drawn on layer 0. Normally when a WBLOCK or BLOCK is inserted it will be put on the current layer in the receiving drawing. However, it will still retain its own layer information for the entities making up the block. Thus if a block was originally created on the FITTINGS layer its entities will always be inserted on that layer. If the receiving drawing doesn't already have a layer called FITTINGS one will be created automatically. The one exception to this is a block created on layer 0. Such blocks and their component entities will always be inserted onto the receiving drawing's current layer.

Inserting a global block is much the same process as that described above. Use the INSERT command and give the WBLOCK name, ie "WDOOR", as the block name and proceed as before. You don't include the ".DWG" extension in the block name. As long as there is no block called "WDOOR" already in the drawing AutoCAD will search the current DOS directory for the drawing "WDOOR.DWG".

As a corollary to WBLOCK insertion, any AutoCAD drawing file can be INSERTed in another drawing. You simply give the drawing name at the "Block name:" prompt and carry on as usual. The insertion base point will be the origin of the external drawing unless a different base point has been specified. When creating a drawing to be used later as a block you can set a suitable base point with the BASE command. For example, to set the point (100,100) as the insertion base point of a drawing you would use the following command sequence:

Command: **BASE** Base point <0.0000,0.0000>: **100,100**

To set the base point back to the origin use the command once more.

Command: **BASE** Base point <100.0000,100.0000>: **0,0**

Making a library of useful symbols

The main items in a small modern fitted kitchen are the sink or basin, the cooker, refrigerator or fridge-freezer, washing machine, dishwasher, storage units and worktops. There are of course many more items that could be included but those mentioned will suffice to illustrate the different block definition methods and also some new editing commands.

Before starting on the rest of the symbols, delete all the doors that you inserted above and change to the FITTINGS layer.

Command: **ERASE**
Select objects: **Window**
First corner: **0,0**
Other corner: **6500,4500**
3 found.
Select objects: <**ENTER**>
Command: **LAYER** or pick the layer pull down list fron the toolbar
?/Make/Set/.../: **S**
New current layer <0>: **FITTINGS**
?/Make/Set/...: <**ENTER**>

Note that in the selection, 3 objects were found. Each block is considered by AutoCAD as a single object.

The kitchen sink

To draw the double drainer sink shown in Figure 6.7 use the LINE command. A rectangular **ARRAY** can be used to draw the drainers. The curved corners of the basin are formed by filleting the rectangle. You might find it useful to zoom in using a window from (800,800) to approximately (2800,2200).

Command: **LINE**
From point: **1000,1000**   (A)
To point: **@1500,0**   (B)
To point: **@0,600**   (C)
To point: **@-1500,0**   (D)
To point: **CLOSE**

The basin is another rectangle.

Command: <**ENTER**>
LINE From point: **1500,1100**   (E)
To point: **@500,0**   (F)
To point: **@0,400**   (G)
To point: **@-500,0**   (H)
To point: **CLOSE**

And the drainer is made up of parallel lines.

Command: <**ENTER**>
LINE From point: **1100,1100**   (J)
To point: **@300,0**   (K)
To point: <**ENTER**>

The sink is taking shape and should look like Figure 6.7(a). The rest of the drainer can be made by using a 5 row by 2 column array as shown in Figure 6.7(b).

Command: **ARRAY**
Select objects: **LAST**

This picks up the last line JK.

Select objects: **<ENTER>**
Rectangular or Polar array (R/P): **R**
Number of rows (---) <1>: **5**
Number of columns (| | |) <1>: **2**
Unit cell or distance between rows (---): **100**
Distance between columns (| | |): **1000**

Now to round off the corners on the basin use the FILLET command. With this command you can replace the sharp intersections between lines and other entities with a circular arc with a given radius. The arc will be drawn so as to be tangential to both intersecting objects. **FILLET** can be picked from the **Construct** pull-down menu. Fillet can also be executed by picking the button from the tool box shown in Figure 6.7.

Command: **FILLET**
Polyline/Radius/<Select two objects>: **RADIUS**
Enter fillet radius <0.0000>: **50**

This sets the radius for the fillet arcs. You only have to set the radius once and it will remain at that value until it is changed by another FILLET command. To operate on the lines pick **Construct/FILLET** once more or press **<ENTER>**. When picking the intersecting lines with the mouse try to pick points near the intersection (see Figure 6.7). With some versions of the menu, the second execution of FILLET starts automatically.

Command: **FILLET**
Polyline/Radius/<Select two objects>: Pick line HG near point H and then pick HE near H.

Now repeat this for the other three corners (points E, F and G). Finally, add the drainage hole using a circle.

Command: **CIRCLE**
3P/2P/TTR/<Center point>: **1750,1400**
Diameter/<Radius>: **50**

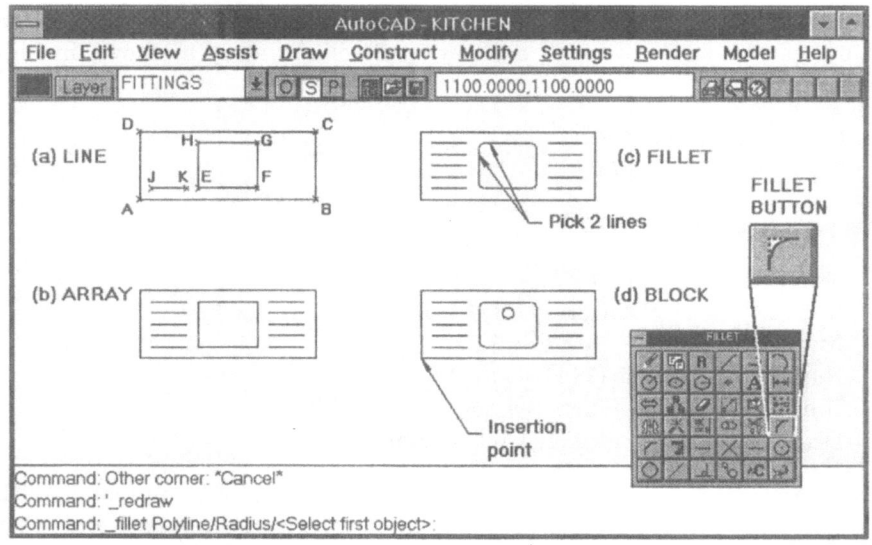

Figure 6.7 Filleting the sink

This completes the double drainer sink. All that remains is to make it into a block. Use the lower left corner as the insertion base point. When everything has been selected, exit from the "select objects" procedure by pressing <**ENTER**> once more.

Command: **BLOCK**
Block name (or ?): **SINK**
Insertion base point: **1000,1000**
Select objects: **Window**
First corner: **1000,1000**
Other corner: **@1500,600**
23 found.
Select objects: <**ENTER**>

Colored blocks

As blocks are inserted onto the layer names that they were created on, they will take up the color of that layer. This means that if you change the color of layer FITTINGS to blue and insert the SINK block it will be drawn blue. In many instances symbols or blocks have meaningful colors which we would like to keep constant, irrespective of the color setting of the layer. In this section a

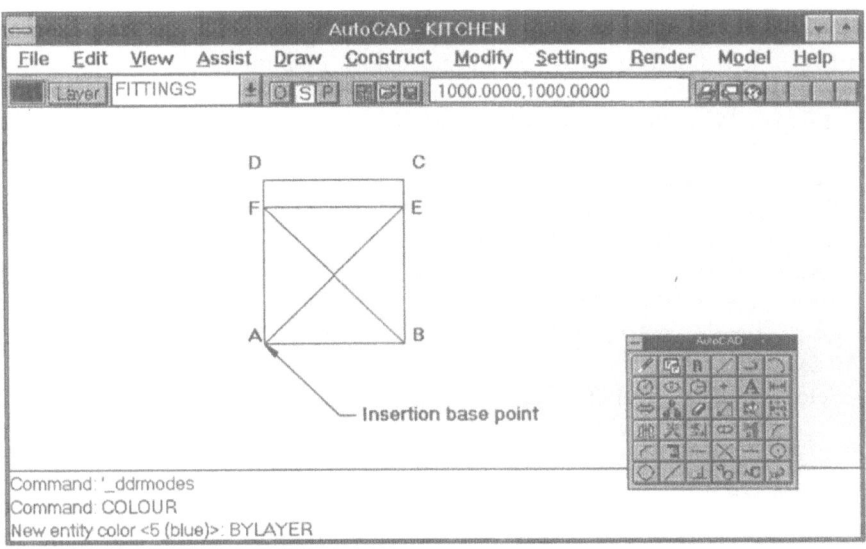

Figure 6.8 Fridge-freezer block

fridge-freezer block (Figure 6.8) will be defined to have a constant blue color and a cooker block will be partly defined as red in the following section.

The color of an AutoCAD entity can be defined by its layer or may be specially set using the COLOR command. Set the entity color to blue and draw the fridge block. Either type the command or pick the color box at the left end of the toolbar and use the dialogue box as in Chapter **3**.

Command: **COLOR**
New entity color: **BLUE**
Command: **LINE**
From point: **1000,1000**   (A)
To point: **@500,0**   (B)
To point: **@0,600**   (C)
To point: **@-500,0**   (D)
To point: **@0,-600**   (A)
To point: **@500,500**   (E)
To point: **@-500,0**   (F)
To point: **@500,-500**   (B)
To point: **<ENTER>**
Command: **COLOR**
New entity color: **BYLAYER**

This completes the fridge-freezer in blue. Always reset the color back to "BY-LAYER" when finished with the special color. This means that new entities will be drawn in the default color assigned to the layer.

Now to group the entities into a block.

Command: **BLOCK**
Block name (or ?): **FFREEZER**
Insertion base point: **1000,1000**
Select objects: **Window**
First corner: **1000,1000**
Other corner: **@500,600**
9 found.
Select objects: **<ENTER>**

Editing a block

To draw the cooker in Figure 6.9 use the normal color for the outline and red circles for the cooking elements. The array command can be used to copy the elements. For convenience you can edit the fridge-freezer block. To do this you will have to insert the block and break it down into its constituent entities. The EXPLODE command does just that.

Command: **INSERT**
Block name (or ?) <DOOR>: **FFREEZER**
Insertion point: **1000,1000**
X scale factor <1>/Corner/XYZ: **<ENTER>**
Y scale factor (default=X): **<ENTER>**
Rotation angle <0>: **<ENTER>**
Command: **EXPLODE** or select if from **Modify** pull-down menu.
Select block reference, polyline, dimension or mesh: **LAST**

EXPLODE only works on blocks that have equal X and Y scale factors. It can be used to break up polylines as well.

Now use the change command to set the colors to "bylayer" and then erase the diagonal lines.

Command: **CHANGE** or pick the CHANGE button from the tool box
Select objects: **Window**
First corner: **1000,1000**
Other corner: **@500,600**
9 found.
Select objects: **<ENTER>**
Properties/<Change point>: **P**

Change what property (Color/Elev/LAyer/LType/Thickness)? **C**
New color <5 (blue)>: **BYLAYER**
Change what property (Color/Elev/LAyer/LType/Thickness)?
  <ENTER>
Command: **ERASE**
Select objects: **1100,1100**   (One of the diagonal lines)
1 selected, 1 found.
Select objects: **1400,1100**   (The other diagonal line)
1 selected, 1 found.
Select objects: **<ENTER>**

Now set the color red and draw the cooking rings.

Command: **COLOR**
New entity color <BYLAYER>: **RED**
Command: **CIRCLE**
3P/2P/TTR/<Center point>: **1150,1150**
Diameter/<Radius>: **75**
Command: **ARRAY**
Select objects: **LAST** 1 found
Select objects: **<ENTER>**
Rectangular or Polar array (R/P): **R**
Number of rows (---) <1>: **2**
Number of columns (| | |) <1>: **2**
Unit cell or distance between rows (---): **200**
Distance between columns (| | |): **200**

This should now look like the picture in Figure 6.9. The only way to edit part
of a block is to decompose it completely into the original entities. One way of
achieving this is to EXPLODE it as above. A similar effect results by prefixing
the block name with an asterisk when inserting it. For example:

Command: INSERT
Block name (or ?) <DOOR>: *FFREEZER
Insertion point: 1000,1000
Scale factor <1>: <ENTER>
Rotation angle <0>: <ENTER>

This gives the same result as the INSERT and EXPLODE used above. This
method can only be used when actually inserting a block. EXPLODE can be
used on a block at any time after it has been inserted. There is also an EX-
PLODE option in the Insert dialogue box in the lower left corner (Figure 6.6).
    Now to block the cooker and reset the color to bylayer.

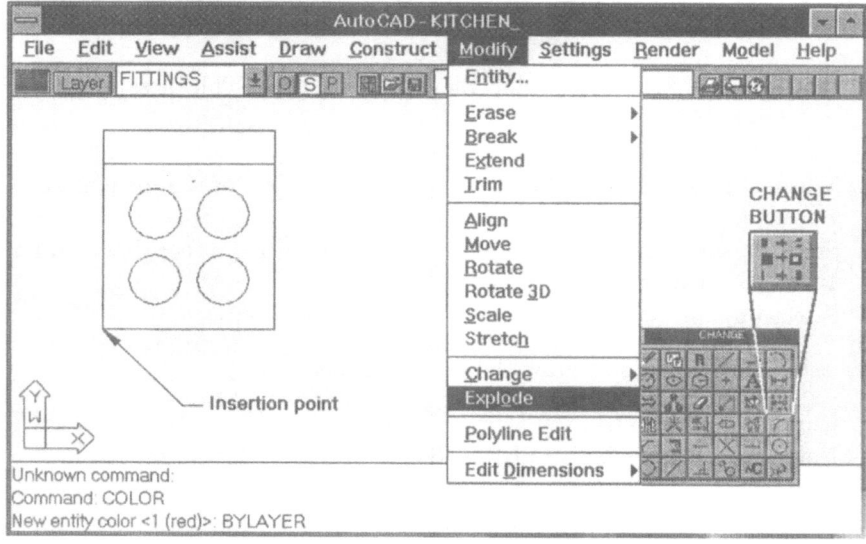

Figure 6.9 Fridge-freezer block

Command: BLOCK
Block name (or ?): COOKER
Insertion base point: 1000,1000
Select objects: Window
First corner: @
Other corner: @500,600
11 found.
Select objects: <ENTER>
Command: COLOR
New entity color <1 (red)>: BYLAYER

One must be careful when defining special colors for blocks or parts of blocks. As blocks can be nested within other blocks it can become difficult to keep track of the colors. The best policy is to keep the coloring of blocks as simple as possible and to use special color settings sparingly.

There is one other color setting that has not been discussed so far. For blocks that are made up of entities on different layers, the BYBLOCK color option can be used. This forces all the entities making up the block to have the same color (whatever color the block is set to). This is irrespective of the color settings of the individual layer and is the converse of having a multi-color block on one layer.

Finally, all of the rules given above for the COLOR command can be equally applied to the LINETYPE command. LINETYPE can be used to override the Ltype settings on individual layers. It works in exactly the same way as COLOR.

Assigning text information to blocks

One of the strongest reasons for using blocks in AutoCAD drawings is that extra non-graphic information can be assigned to the blocks. Furthermore, the attribute information attached to a block can be varied each time the block is inserted. All this information can be gathered together and written to an external file which can then be transferred to a bill of materials program to extract quantities and cost estimates.

In this section a simple attribute will be defined for an electrical appliance block. This will identify what type of appliance has been inserted. Attributes are also defined for cupboard units and worktop finishes.

Defining an attribute

An attribute is treated by AutoCAD like any other drawing entity. Once it has been defined it can be included with other entities to form a block. To create the electrical appliance block (Figure 6.11), first draw its outline.

```
Command: LINE
From point: 1000,1000
To point: @600,0
To point: @0,500
To point: @-600,0
To point: close
```

Now include hot and cold inlet pipes and the waste outlet point at the back of the box.

```
Command: LINE
From point: 1100,1500
To point: @0,100
To point: <ENTER>
Command: <ENTER>
LINE From point: 1200,1500
To point: @0,100
To point: <ENTER>
Command: <ENTER>
LINE From point: 1300,1500
```

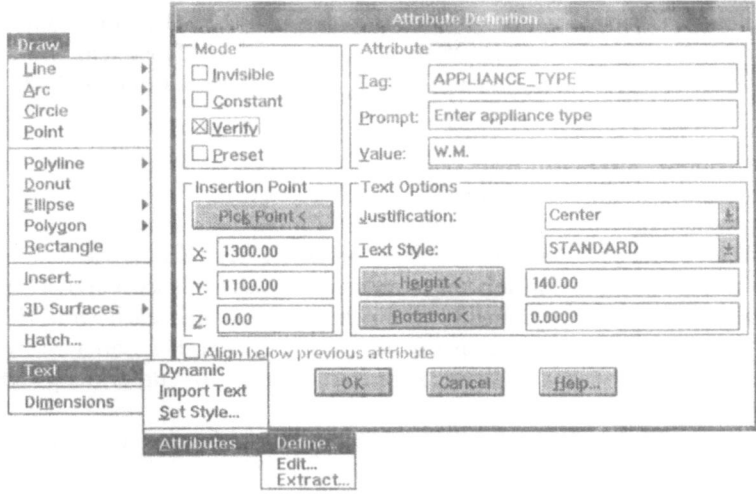

Figure 6.10 Attribute menu and dialogue box

To point: **@0,100**
To point: **<ENTER>**

Now to define the attribute pick **Draw/Text** from the menu bar and then pick **Attributes** and **Define...**, as shown in Figure 6.10. This brings up the Attribute Definition dialogue box.

The dialogue box is divided into four sections, namely, Mode, Insertion point, Attribute and Text options.

In the Mode part of the dialogue box click the "Verify" box to set this mode on. This means that when the block is inserted you will be asked to verify that the value of the attribute is correct. All the other modes are okay for this block. It will not be invisible, constant, or preset but will be verified.

The insertion point **(1300,1100)** is in the lower middle of the rectangle already drawn. If you pick the Pick Point button you can use the mouse to select the point from the graphics screen.

Moving to the Attribute part of the dialogue box, the attribute tag is the name for that attribute and will be used by AutoCAD to identify it. Any characters can be used for the tag except blank spaces. Type **APPLI-ANCE_TYPE** in this field. The prompt is the message that will be displayed when the block is being inserted. It should be clear so that others using your library of blocks can understand what is required of them. The default value will be offered in AutoCAD's usual way. The "W.M." stands for washing machine; the alternative will be "D.W." for dishwasher. You can use more descriptive

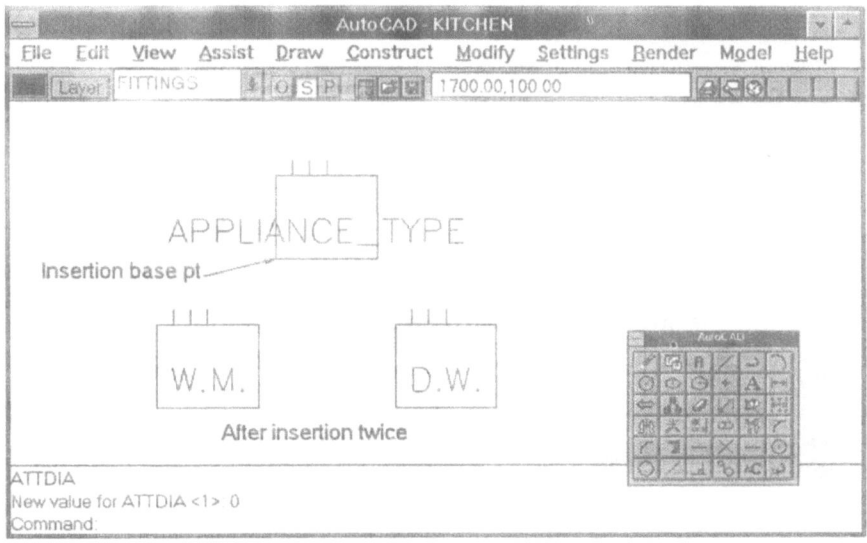

Figure 6.11 Electrical appliance blocks with attributes

attribute values if you wish. The command then asks for the start point of the attribute text in a similar fashion to the TEXT command. Position it centrally in the appliance box and use a large text height to make it visible.

Finally, you must specify how the attribute will be displayed. The text options are similar to those for the standard DTEXT command. Pick the arrow at the end of the Justification field. This gives a pull-down list of the different types of justification. Pick **Center**. Then change the text height to **140**. When your dialogue box matches that in Figure 6.10 pick **OK**.

The text "APPLIANCE_TYPE" should be written across the box as shown in Figure 6.11. This will be replaced by the actual attribute value when the block has been created and inserted. To create the block with the attribute, pick **Construct/Block** or type **BLOCK** at the command prompt. Select the objects for inclusion using a window big enough to surround the attribute as well.

Command: **BLOCK**
Block name (or ?): **APPLIANCE**
Insertion base point: **1000,1000**
Select objects: **Window**
First corner: **300,700**
Other corner: **2400,1900**

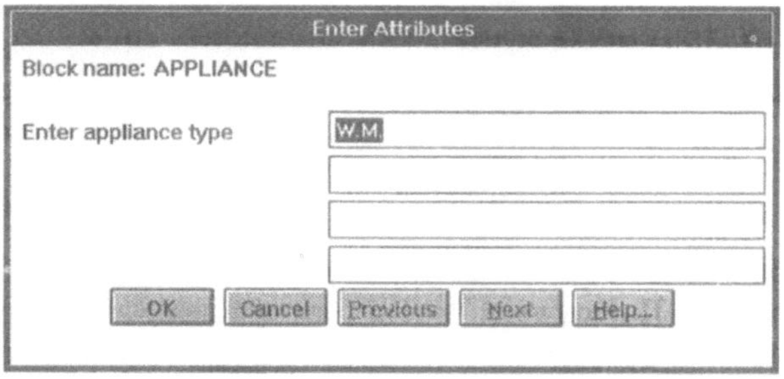

Figure 6.12 Enter Attributes dialogue box

8 found.
Select objects: <ENTER>

Now to check that it works insert it at (1000,1000). If you are confronted
with the Attribute Dialogue Box, similar to the one shown in Figure 6.12, just
pick the **OK** box to proceed. To disable the dialogue box reset the system
variable, ATTDIA, to 0 by typing the variable name at the command prompt
followed by the desired value (ATTDIA = 1 enables the dialogue box). You
can decide whether you prefer entering the attributes via the dialogue box or
the keyboard. If a block has a number of attributes then the dialogue box is
quite useful. If the block has only one or two attributes then the keyboard is
quicker. Note that the verify mode only applies to keyboard input.

Command: **ATTDIA**
New value for ATTDIA <>: **0**
Command: **INSERT**
Block name (or ?) <FFREEZER>: **APPLIANCE**
Insertion point: **1000,1000**
X scale factor <1>/Corner/XYZ: **<ENTER>**
Y scale factor (default=X): **<ENTER>**
Rotation angle <0>: **<ENTER>**
Enter attribute values
Enter appliance type <W.M.>: **<ENTER>**
Verify attribute values
Enter appliance type <W.M.>: **<ENTER>**

The block should now be inserted with the "W.M." written in it. Try inserting
it again but give "D.W." as the attribute value. You should get something like
Figure 6.11. Using attributes in this way saves having to define different blocks
for each type of large electrical appliance.

The attributes of a block can be invisible. This is useful when the text is not relevant to the actual picture or if it would crowd the drawing too much. In the following sequence you will define a block for the cupboard units with an invisible attribute giving information on the number of doors on the unit. As blocks can have more than one attribute you can also include an attribute for the type of finish the customer requires on the unit.

The standard unit size is 500mm wide by 600mm deep. Two units can be joined to form a double, etc. Draw the unit using a polyline this time and use a thicker line to indicate the side with the door. Be sure to **ERASE** all the copies of the block you have just inserted.

Command: **PLINE**
From point: **1000,1000**
Current line-width is 0.0000
Arc/Close/.../<Endpoint of line>: **@600<90**
Arc/Close/.../<Endpoint of line>: **@500,0**
Arc/Close/.../<Endpoint of line>: **@0,-600**
Arc/.../Width/<Endpoint of line>: **Width**
Starting width <0.0000>: **25**
Ending width <25.0000>: **<ENTER>**
Arc/Close/.../<Endpoint of line>: **CLOSE**

Now define the first attribute for the number of doors in the unit. Pick **Draw/Text/Attributes/Define** from the menu bar. Then click **Invisible** and **Verify**. The insertion point is **(1250,1400)**. Type **DOORS** for the attribute tag, the prompt as shown and give a default value of **1**. The text options are as before. When your display matches Figure 6.13 pick **OK**.

Now repeat this for the surface finish attribute. An alternative method is to use the ATTDEF command rather than the dialogue box.

Command: **ATTDEF**
Attribute modes – Invisible:Y Constant:N Verify:Y Preset:N
Enter (ICVP) to change, RETURN when done: **<ENTER>**
Attribute tag: **FINISH**
Attribute prompt: **Enter the type of surface finish**
Default attribute value: **OAK**
Justify/Style <Start point>: **J**
Align/Center/...: **C**
Center point: **1250,1100**
Height <0.20>: **140**
Rotation angle <0>: **<ENTER>**

You are ready to make the block now.

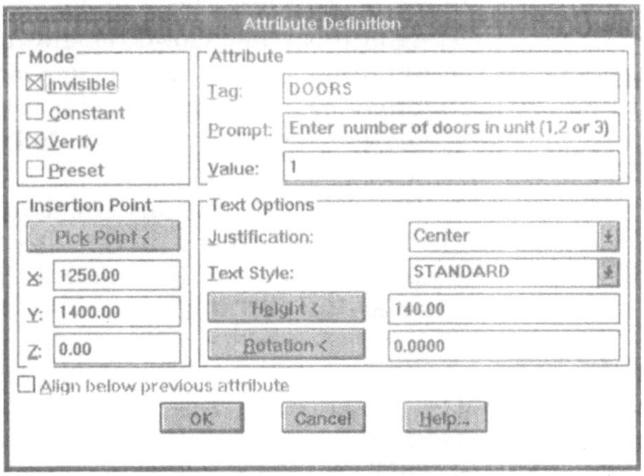

Figure 6.13 Cupboard attributes

Command: **BLOCK**

Block name (or ?): **CUPBOARD**

Insertion base point: **1000,1000**

Select objects: **Window**

First corner: **300,300**

Other corner: **1900,1900**

3 found.

Select objects: **<ENTER>**

When you insert this block later you will be prompted for the two attributes. The order will be the reverse of the definition, ie you will be prompted for the surface finish first. The block will be drawn with both attributes invisible (Figure 6.14). You can experiment with inserting this block or wait for the next drawing in which all the blocks will be used to assemble the AutoCAD Express Fitted Kitchen. Be sure to **ERASE** everything before proceeding.

To get a listing of all the blocks in the drawing you can use the "?" option with either the BLOCK or INSERT commands. The display will flip to text mode and give the names. Alternatively, pick **Draw/Insert** from the menu bar. Then pick the **BLOCK** button in the Insert dialogue box. You can then pick the desired block from the list by double clicking it followed by **OK** (Figure 6.15).

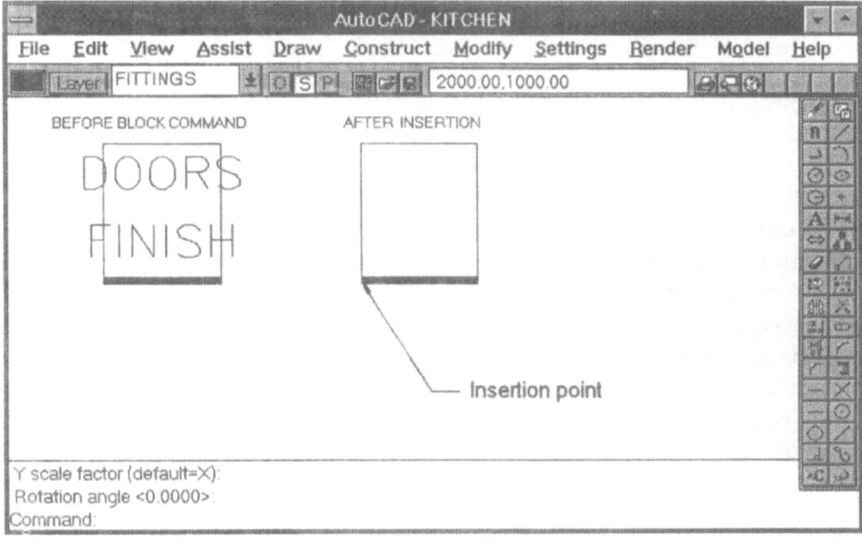

Figure 6.14 The cupboard with two attributes

Figure 6.15 List of defined blocks

Drawing the kitchen

Before inserting all the blocks, let's do the outline of the kitchen. It's only a small kitchen with a simple shape. This should be done on layer 0.

```
Command: LAYER or use the tool bar
?/Make/Set/...: S
New current layer <FITTINGS>: 0
?/Make/Set/...: <ENTER>
Command: PLINE
From point: 1000,1000                                            (A)
Current line-width is 0.0000
Arc/.../Width/<Endpoint of line>: Width
Starting width <0.0000>: 10
Ending width <10.0000>: <ENTER>
Arc/Close/.../<Endpoint of line>: @3300,0                        (B)
Arc/Close/.../<Endpoint of line>: @0,3100                        (C)
Arc/Close/.../<Endpoint of line>: @-3300,0                       (D)
Arc/Close/.../<Endpoint of line>: CLOSE
```

Now to draw the outside edge of the wall you can use AutoCAD's OFFSET command. This can be used to draw objects which are the same as the original but offset by a given amount. For example, it can be used to draw parallel lines. The offset command causes offset circles to be concentric with the original and has a similar effect on closed polylines (Figure 6.16). Pick **Construct/Offset** from the menu bar. You will then be asked for an offset distance, which will be the thickness of the wall. Then you must select the object to offset by picking it or giving a point on it. Finally you must choose on which side of the original to place the offset.

```
Command: OFFSET
Offset distance or Through <Through>: 100
Select object to offset: 1000,1000                              (A)
Side to offset: 900,1000                                        (E)
```

This picks a point on the outside of the original rectangle. AutoCAD then asks for another object to offset. Press <ENTER> to exit the command.

```
Select object to offset: <ENTER>
```

Note that you cannot use Window, crossing or last to select the object.

Now that the walls are up we must break through some door openings. This introduces another AutoCAD editing command, BREAK. Before this, Zoom in for a closer look.

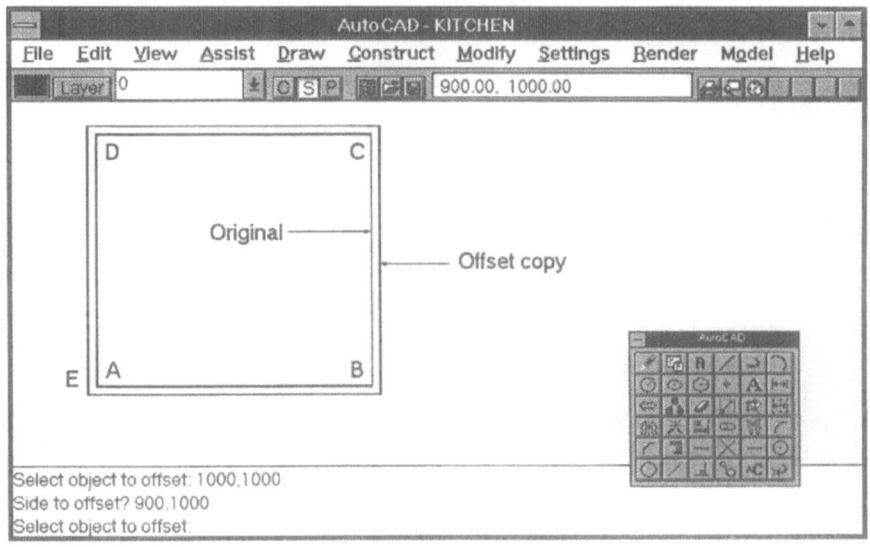

Figure 6.16 Offsetting a closed polyline

Command: **ZOOM**
All/.../Window/<Scale(X/XP)>: **W**
First corner: **500,500**
Other corner: **3000,2400**

On the Break sub-menu there are three options (Figure 6.17). The first is
"Select Object, 2nd Point". This breaks the selected entity from the first point
picked to the second. Pick **Modify/Break** and **Select Object,2nd Point**.

Command: BREAK
Select object: **1000,1200**   (F)
Enter second point (or F for first point): **@0,800**   (G)

This breaks an 800mm opening from point F to G (Figure 6.17). The first
point had the dual purpose of selecting the object and indicating the start of
the break. Sometimes it is not convenient to select the object at the actual
break point. In such a case you would use the "F" reply at the second prompt
above. Or you can pick **Select Object, Two Points** from the sub-menu. The
command line will echo the following:

Command: BREAK
Select object: **900,1600**   (A point between H and J)

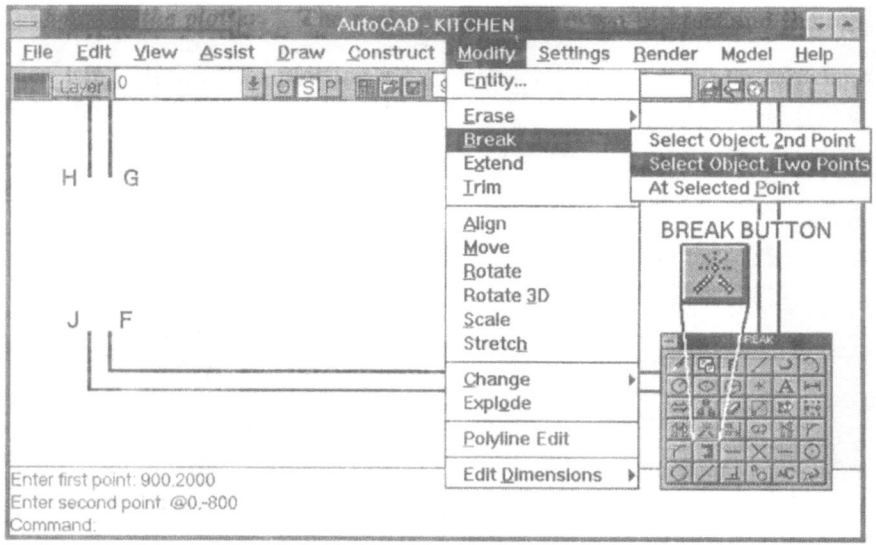

Figure 6.17 Breaking a polyline

Enter second point (or F for first point): **F**
Enter first point: **900,2000**   (H)
Enter second point: **@0,-800**   (J)

This achieves a similar result and allows you extra flexibility in selecting the object. This can be particularly useful when two objects intersect at the desired break point. AutoCAD may select the wrong object if you pick the intersection point. BREAK can also be used on lines, circles and arcs.

The final stage in making the opening is to draw the two short polylines.

Command: **PLINE**
From point: **900,1200**   (J)
Current line-width is 10.0000
Arc/Close/.../<Endpoint of line>: **@100,0**   (F)
Arc/Close/.../<Endpoint of line>: **<ENTER>**
Command: **<ENTER>**
PLINE
From point: **900,2000**   (H)
Arc/Close/.../<Endpoint of line>: **@100,0**   (G)
Arc/Close/.../<Endpoint of line>: **<ENTER>**

ZOOM to the opposite corner and insert another door opening and a window. Break the door from (4400,2700) to (4400,3500) and draw the window from (1600,4150) to (3600,4150). Repeat the breaks for the inside polyline and then join up all the loose ends (Figure 6.18). This time pick the Break button from the tool box or type the command. This is the same as "Select Object,2nd Point".

Command: **ZOOM**
All/.../Window/<Scale(X/XP)>: **W**
First corner: **1500,2000**
Other corner: **4600,4400**
Command: **BREAK** or pick the Break button
Select object: **4400,2700**   (K)
Enter second point (or F for first point): **@0,800**   (L)
Command: **<ENTER>**
BREAK Select object: **4300,2700**   (M)
Enter second point (or F for first point): **@0,800**   (N)

To join up the loose ends draw four more polylines and draw the window PQRS.

Command: **PLINE**
From point: **4300,2700**   (M)
Current line-width is 10.0000
Arc/Close/.../<Endpoint of line>: **@100,0**   (K)
Arc/Close/.../<Endpoint of line>: **<ENTER>**
Command: **<ENTER>**
PLINE
From point: **4300,3500**   (N)
Current line-width is 10.0000
Arc/Close/.../<Endpoint of line>: **@100,0**   (L)
Arc/Close/.../<Endpoint of line>: **<ENTER>**
Command: **<ENTER>**
PLINE
From point: **1600,4100**   (P)
Current line-width is 10.0000
Arc/Close/.../<Endpoint of line>: **@0,100**   (Q)
Arc/Close/.../<Endpoint of line>: **<ENTER>**
Command: **<ENTER>**
PLINE
From point: **3600,4100**   (R)
Current line-width is 10.0000
Arc/Close/.../<Endpoint of line>: **@0,100**   (S)
Arc/Close/.../<Endpoint of line>: **<ENTER>**

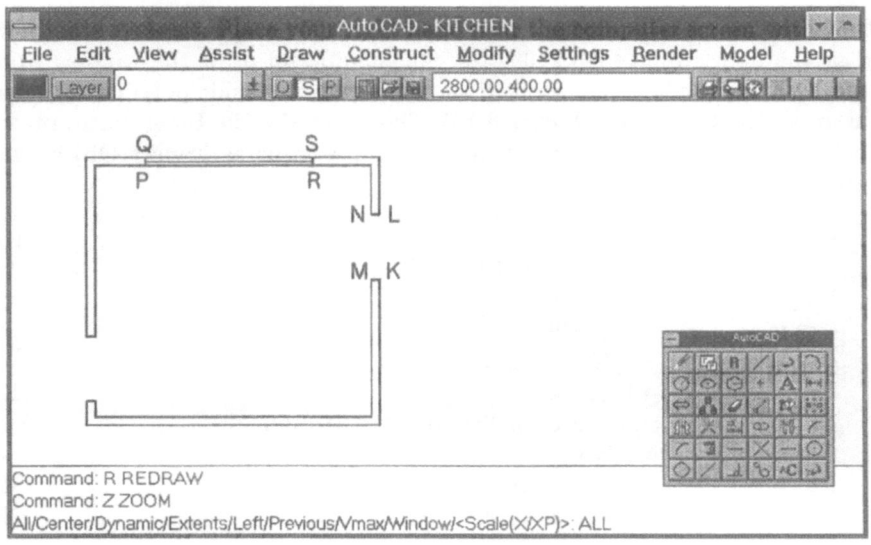

Figure 6.18 The empty kitchen

Now draw the window glazing as a horizontal line.

Command: **LINE**
From point: **1600,4150**   (Mid pt of line PQ.)
To point: **@2000,0**   (Mid pt of line RS.)
To point: **<ENTER>**
Command: **ZOOM**
All/.../Window/<Scale(X/XP)>: **A**

Assembling the fitted kitchen

Assembling the fittings and fixtures of the kitchen is now just a matter of inserting all the blocks in their correct locations. If the customer wants things moved around then that's no problem with AutoCAD. Change layer to FIT-TINGS before beginning the task.

Insert the sink first with a view out the window. Pick **Draw/Insert**, then pick the **Block...** button followed by **SINK**. Complete the dialogue box as shown in Figure 6.19. Disable the "Specify Parameters on Screen" and input the coordinates of the insertion point, T, **(1600,3500)** as shown. Use the default scales and pick **OK**.

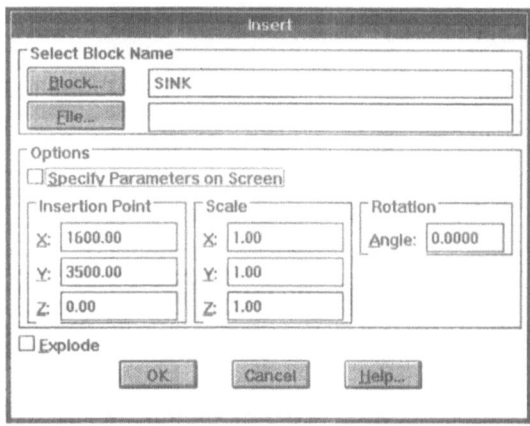

Figure 6.19 Inserting the sink

Now put the washing machine and dishwasher beside the sink (Figure 6.20). In the following sequence, ATTDIA is 0 and keyboard input is used.

Command: **INSERT**
Block name (or ?) <SINK>: **APPLIANCE**
Insertion point: **3100,3500**   (U)
X scale factor <1>/Corner/XYZ: **<ENTER>**
Y scale factor (default=X): **<ENTER>**
Rotation angle <0>: **<ENTER>**
Enter appliance type <W.M.>: **<ENTER>**
Verify attribute values
Enter appliance type <W.M.>: **<ENTER>**
Command: **INSERT**
Block name (or ?) <APPLIANCE>: **<ENTER>**
Insertion point: **3700,3500**   (V)
X scale factor <1>/Corner/XYZ: **<ENTER>**
Y scale factor (default=X): **<ENTER>**
Rotation angle <0>: **<ENTER>**
Enter appliance type <W.M.>: **D.W.**
Verify attribute values
Enter appliance type <D.W.>: **<ENTER>**

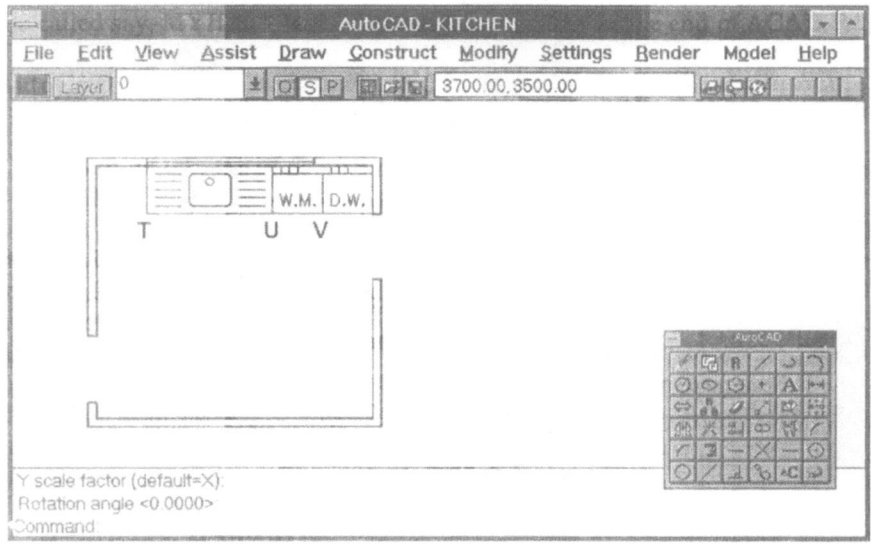

Figure 6.20 Appliances in position

The rest of the block information is given in tabular form. Use the INSERT command and the appropriate responses taken from Table 6.2. Figure 6.21 shows the fitted kitchen.

Editing attributes

The first thing to do now is to check that all the attributes on the drawing are in fact correct. To make the invisible attributes appear on the drawing use the **ATTDISP** command. This command allows you to alter the display setting for all the attributes in the drawing. The normal display mode is that only attributes defined as visible are shown. Setting ATTDISP to ON causes all attributes to become visible irrespective of their definition. ATTDISP OFF would cause all to become invisible.

Command: **ATTDISP**
Normal/ON/OFF <Normal>: **ON**

From Figure 6.22 we can see that the cupboard in the lower right-hand corner has 2 doors. Since it is in the corner, one of these doors will be blocked by the other cupboard. To save the customer unnecessary expense only one

Table 6.2 Block insertion parameters

Block name	Insertion point	X scale	Y scale	Rotation	Attribute values
FFREEZER	3700,2600	1	1	−90	
COOKER	1600,2600	1	1	90	
CUPBOARD	1600,3100	2	1	90	PINE, 1
CUPBOARD	1600,2100	1	1	90	PINE, 1
CUPBOARD	3700,2100	2.2	1	−90	PINE, 2
CUPBOARD	3700,1600	3	1	180	PINE, 3
DOOR	950,2000	1	1	−90	
DOOR	4350,2700	−1	1	−90	

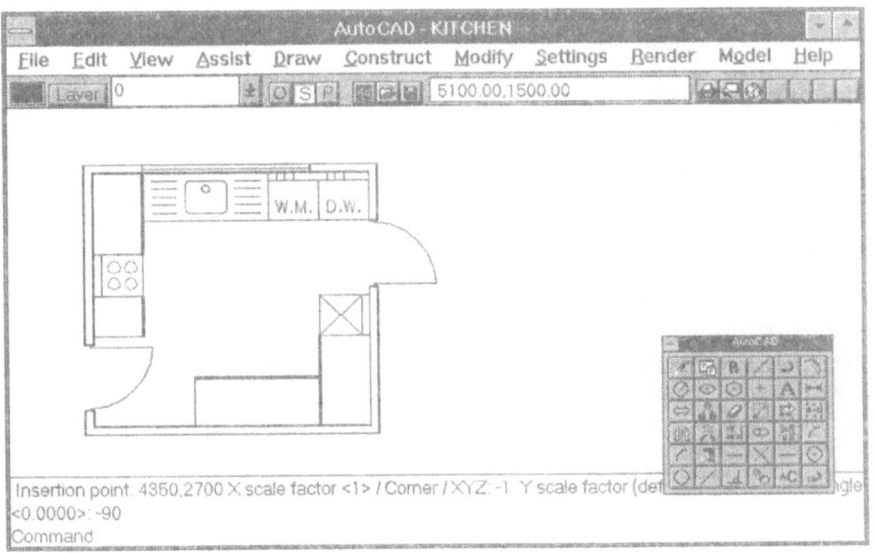

Figure 6.21 The fitted kitchen

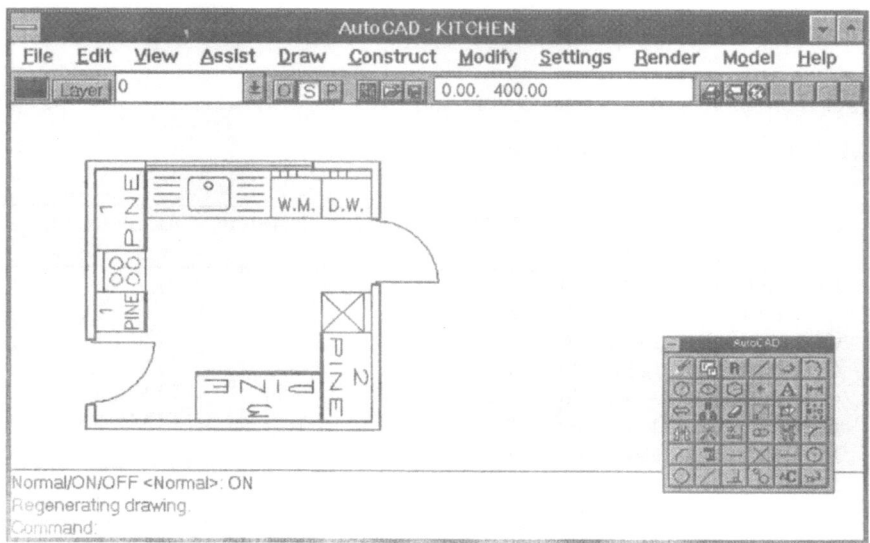

Figure 6.22 ATTDISP set ON

door should be provided. Thus the attribute must be edited. The easiest way
to do this is to use the DDATTE dialogue box.

Pick **Draw/Text/Attributes** from the menu bar. Then pick **Edit....**
You are then prompted to select the block to edit.

Command: _ddatte

<Select Text or Attdef object>/Undo: **3700,1900**

The dialogue box shown in Figure 6.23 should then appear. This gives the two
attribute values and their prompt. To change the 2 doors to 1 move the arrow
cursor and click in the rectangle containing the 2. Then type 1. This should
overwrite the old value. You may have to use the backspace key to delete the
"2" before typing the 1. To execute this change pick the **OK** box at the bottom
of the dialogue screen. When the dialogue box disappears the "2" in the block
will be changed to "1".

To set the attribute display back to normal use ATTDISP once more.

Command: **ATTDISP**

Normal/ON/OFF <ON>: **Normal**

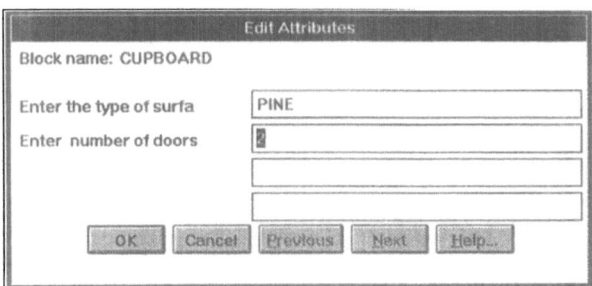

Figure 6.23 Dialogue box for attribute edits

A simple bill of materials

Information about all blocks that have attributes can be extracted from the drawing and output to a text file. The text file could then be incorporated in a report, a spreadsheet or a bill of quantities. There are three ways to extract this information, using AutoCAD's DXF format, a CDF file or an SDF. The easiest way is to use a template file to control which attributes are required for output to either an SDF or a CDF file.

A CDF file (Comma Delimited File) is an ASCII text file where the block's information fields are separated by commas. In an SDF they are separated by spaces while the DXF file uses AutoCAD's drawing exchange protocol. The latter is of use to program developers but is rather cumbersome.

In order to use the SDF or CDF method you first have to define a template file. This has to be an ASCII file created with a text editor. The file name must have the extension ".TXT". The file given below is suitable for use on the KITCHEN drawing. The comments in brackets should *not* be included in the actual KITCHEXT.TXT file. The first item on each line is a key-word indicating what is to be extracted, the second gives the format for writing that information to the CDF or SDF file. If the information is a number then the second field should start with "N"; if it is text then "C" should be used. The first three digits after the "C" or "N" indicate how many spaces are to be reserved for that field. The second three indicate how many digits are required after the decimal point. Thus "N007001" would output a number such as "1234.5" but not "1234.56". Always leave one space for a possible minus sign. Make sure that there are no blank lines in the template file.

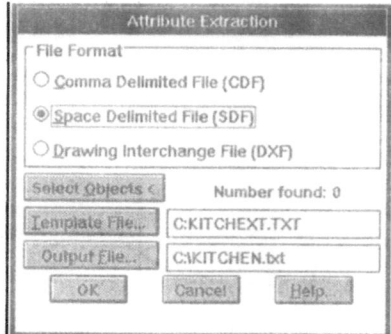

Figure 6.24 Attribute extraction dialogue box

BL:NAME	C010000	(Block name with up to 10 characters)
FINISH	C007000	(Finish attribute with up to 7 characters)
APPLIANCE_TYPE	C007000	(Appl. type attribute with up to 7 characters)
DOORS	N006000	(No. of doors attribute up to 6 digit integer)
BL:XSCALE	N007001	(X scale factor, number with 1 digit after ".")
BL:YSCALE	N007001	(Y scale factor, number with 1 digit after ".")

KITCHEXT.TXT

When you are in the drawing editor with the KITCHEN drawing you can generate the bill of materials using the DDATTEXT command. Pick **Draw/Text**, then **Attributes/ Extract**.... The dialogue box shown in Figure 6.24 will then appear. Click the **SDF** format and give the template and output files as shown.

AutoCAD should echo "6 records extracted". Now to see the result use the external command TYPE.

Command: **TYPE**
File to list: KITCHEN.TXT

APPLIANCE		W.M.	0	1.0	1.0
APPLIANCE		D.W.	0	1.0	1.0
CUPBOARD	pine		1	1.0	1.0
CUPBOARD	pine		1	2.0	1.0
CUPBOARD	pine		1	2.2	1.0
CUPBOARD	pine		3	3.0	1.0

KITCHEN.TXT

Only blocks that contain attributes included in the template file are extracted. The DOOR, SINK COOKER and FFREEZER block don't appear as they have none of the attributes. You can include more information fields, if desired, to extract the insertion points and other data. See the AutoCAD Reference Manual for the full list of key-words. The SDF file is suitable for use with a language FORTRAN and some database programs while the CDF format works with BASIC.

To finish this session pick **File/Exit AutoCAD** followed by **Save Changes** if you are prompted to do so. Outside of AutoCAD you can use your text editor or spreadsheet to manipulate the file Kitchen.txt.

Hints on using blocks

Blocks should be employed for commonly used shapes and symbols. You can build up a library of symbols in the form of WBLOCKS. This is more efficient than using a default drawing that contains many block definitions.

Try to develop a consistent method for naming your blocks and store the WBLOCKS in a separate DOS directory. For example you might create a subdirectory of \ACAD called "\ACAD \SYMBOLS" and include that path in the WBLOCK file name. The more symbols you have the more effort you will have to devote to managing this storage.

Blocks forming part of a symbols library should always contain some attribute definitions. This will enable information about them to be extracted for bill of material purposes.

If a complicated object (containing more than 20 entities) appears more than once in a drawing then BLOCK it and INSERT it as required. This is more efficient than simply copying all the entities, as AutoCAD stores the shape of the block only once. The only other items of information required are the relevant insertion points. This saves memory and disk space.

Avoid using nested blocks if possible. By nested, one means that one block can contain others. This can lead to difficulties if you want to make changes to the block definition at a later date.

AutoCAD's inquiry commands

A number of other information extraction commands are available in the Assist/Inquiry pull-down menu. These allow you to get the area enclosed by a polyline (**AREA** command), the distance between two points (**DIST** command) or the coordinates of a particular point (**ID**). **LIST** allows you to find out all the information about any entities. This is helpful for finding out which layer an entity has been drawn on. The **STATUS** command gives a listing of the current drawing status and also the amount of memory available in the computer and on the disk. This is useful for keeping tabs on your hardware and the current AutoCAD default settings.

Summary

This chapter has introduced the fundamentals of AutoCAD block creation. Blocks are a useful feature for storing standard symbols. They can be exported from the original drawing and so be made available to other drawings. Some more editing features have been used to create the blocks. Finally, the information extraction facility has been demonstrated.

You should now be able to:

Create blocks and global blocks.
List the names of blocks in a drawing.
Assign text information to blocks and edit it.
Round off corners with the FILLET command.
Offset objects and break gaps in entities.
Alter the entity color.
Interrogate the drawing database.

Chapter 7 ADVANCED DRAWING AND DIMENSIONING

General

Most of AutoCAD's drawing commands have been covered by now. In this chapter you will discover the reason for creating AutoCAD drawings at full scale. The various dimensioning and measuring commands all calculate their distances in the drawing units. To demonstrate the more important automatic dimensioning facilities you will draw a relatively simple object, a mechanical engineer's gland! (Figure 7.1). In doing this a few more new commands will be introduced. Then the dimensions will be added.

Drawing a gland

Start a new AutoCAD drawing and give it the name "GLAND". Set the LIM-ITS from (0,0) to (65,45) and make sure that decimal units are being used with 4 places after the decimal point. You will need the layers given in Table 7.1. To begin the drawing put the center-lines of the circles at convenient locations (Figure 7.2). Use ˆO to turn on the ORTHO mode or pick the O button from the tool bar. If the lines don't appear with dashes change the LTSCALE value to 2.5.

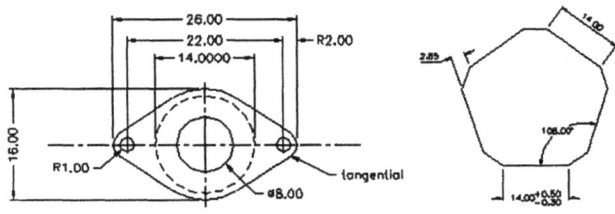

Figure 7.1 Engineer's gland

Table 7.1 Layer settings for drawing GLAND

Layer name	State	Colour	Linetype
0	On	7 (white)	CONTINUOUS
CLINE	On	7 (white)	CENTER
DASH	On	7 (white)	DASHED
DIMENSIONS	On	7 (white)	CONTINUOUS
GLAND	On	7 (white)	CONTINUOUS
POLYGON	On	7 (white)	CONTINUOUS

Current layer: CLINE

```
Command: LINE
From point: 13,20                                    (A)
To point: @44,0                                      (B)
To point: <ENTER>
Command: <ENTER>
LINE From point: 35,11                               (C)
To point: @0,18                                      (D)
To point: <ENTER>
```

POINT and DIVIDE

These are the major axes for the gland's cylinder. The line AB is 44cm long
and the distance from the center of the gland to the bolt holes on either side is
11cm. Thus the quarter points of ab can be used to position the holes. To find
the quarter points use the **Divide** command from the **Construct** pull-down
menu. This can be used to divide a line, arc or polyline into any number of
equal segments. Rather than actually break the line into different entities the
DIVIDE command inserts AutoCAD POINTs at the relevant intervals.

POINTs are drawing entities. Their main use is for marking special locations for object snapping. Object snap "node" jumps to the nearest point
entity. These normally appear on the drawing as dots. To create a point entity
pick **Point** from the **Draw** pull-down menu. You then give it a location and
a dot is drawn. Dots have no dimension and are difficult to see, particularly if
the grid is on. AutoCAD gives a number of options for the display of points in
the Points Style dialogue box (Figure 7.3). Pick **Settings/Point Style...** see
the different styles. Then pick the fourth box on the top row, containing "X".
Then set the point size to 5% relative to the screen size. This relative sizing
means that no matter how far in or out we zoom and magnify the drawing,
the points will appear the same size.

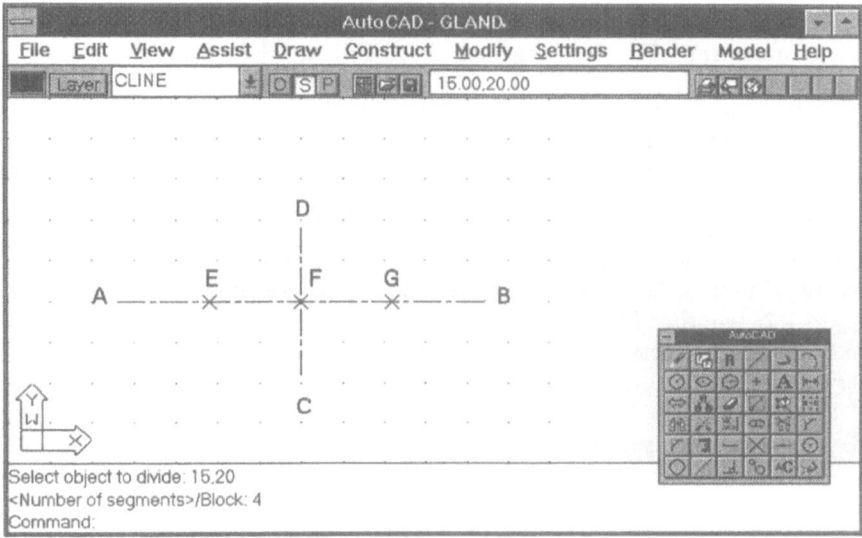

Figure 7.2 Dividing line

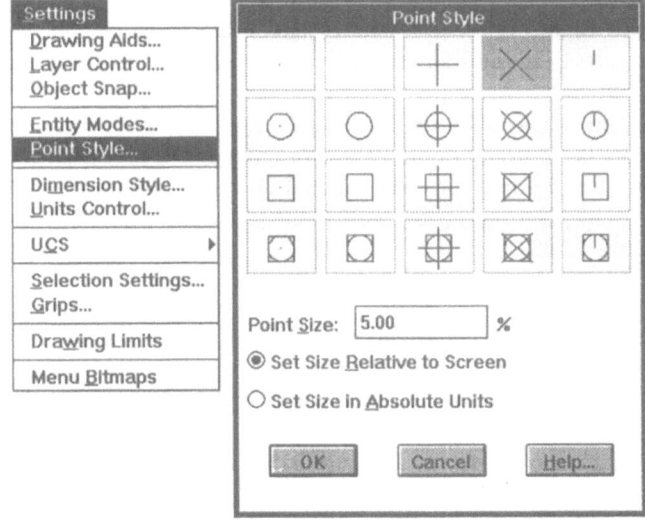

Figure 7.3 Point style dialogue box

Having set the point style it remains to divide the line AB into four. Pick **Construct/Divide** from menu bar. You are then asked for the object to divide. Pick a point on the line AB and give the number of segments as 4.

Command: divide
Select object to divide: **16,20**   (Point on AB)
<Number of segments>/Block: **4**

The "Block" option allows you to insert a named block at the dividing locations instead of points. The points should now appear along the line as shown in Figure 7.2. Another feature of this command is that all the points are put in the "previous" selection set and can be deleted by picking "ERASE" followed by "Previous".

AutoCAD's MEASURE command is very similar to DIVIDE but it puts the point markers at multiples of a specified distance from the end point of the entity. Thus you could use MEASURE to do the same as the above by giving the distance to measure out as 11 units.

Command: MEASURE
Select object to measure: 16,20   (Point on AB)
<Segment length>/Block: 11

This will include as many markers as will fit on the line. With MEASURE four will be drawn at 11, 22, 33 and 44 units from the point A. It is important to pick the line to be measured near to the end you want the measurement to start, particularly if the line length is not a whole multiple of the segment length.

You can now set an Object Snap Running mode to "Node" to snap the circle centers to the dividing points. Pick **Settings/Object Snap...** and then pick **Node** in the dialogue box shown in Figure 7.4. Then pick **OK**. Before drawing the circles, change to layer GLAND and draw circles at the points E, F and G (Figure 7.2).

Command: DDOSNAP
Command: **LAYER** or use the tool bar.
?/Make/Set/...: **S**
New current layer <CLINE>: **GLAND**
?/Make/Set/...: **<ENTER>**

Use diameters of 16, 14, 8, 4 and 2 as given below. Pick **Draw/Circle** from the menu bar and **Center,Diameter** for three large circles (Figure 7.5). The centers will automatically snap to the point entities.

Command: CIRCLE
3P/2P/TTR/<Center point>: (Pick point at F)

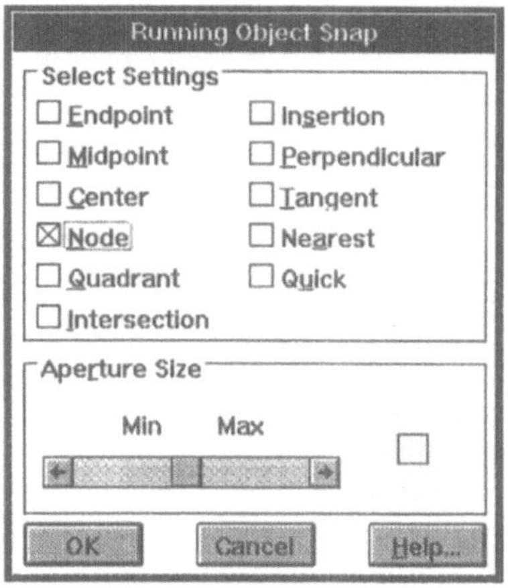

Figure 7.4 Object snap running mode

Diameter/<Radius>: _diameter Diameter: **16**
Command: CIRCLE
3P/2P/TTR/<Center point>: (Pick point at F)
Diameter/<Radius>: _diameter Diameter: **14**
Command: CIRCLE
3P/2P/TTR/<Center point>: (Pick point at F)
Diameter/<Radius>: _diameter Diameter: **8**

Now draw the two small circles at the left hand node. Again use the diameter option.

Command: **CIRCLE**
3P/2P/TTR/<Center point>: (Pick point at E)
Diameter/<Radius>: **D** Diameter: **4**
Command: **CIRCLE**
3P/2P/TTR/<Center point>: (Pick point at E)
Diameter/<Radius>: **D** Diameter: **2**

You will place the objects at G later using a mirroring operation. Now erase the three dividing points and turn off the Object Snap running mode. A quick way to do this is with the OSNAP command.

Command: **ERASE**
Select objects: **Previous**
3 found.
Select objects: **<ENTER>**

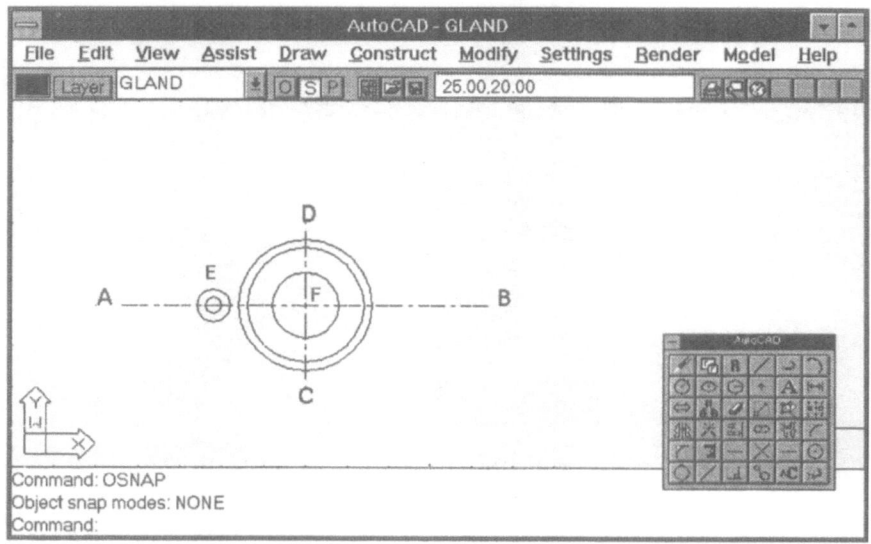

Figure 7.5 Gland circles

Command: **OSNAP**
Object snap modes: **NONE**

If the points are not deleted and you get the message "No previous selection set" it is probably because you have used a command that required you to "Select objects:". If you do such a command after the DIVIDE and before the erase then the selection set will be altered. In that case you will have to delete the points individually.

You can build up a selection set of objects for use with the "previous" option in ERASE and other commands. This is done with the SELECT command from the Edit pull-down menu. The format is just like ERASE but without anything being deleted.

Command: **SELECT**
Select objects: Pick objects or use Window, Crossing etc.
Select objects: **<ENTER>**

To draw the flange for the gland, zoom in on the four circles and draw two lines tangential to both of the outer circles (Figure 7.6). Use **ZOOM Center** to position the center of the large circles at the middle of the screen. A magnification factor of **3** should be suitable.

Command: **ZOOM**

All/Center/.../Window/<Scale(X/XP)>: **C**

Center point: **35,20**   (Center of the large circles)

Magnification or Height <45.0000>: **2X**

Be sure to type the "X" after the "2". This ensures that the circles will be
magnified. Responding with just a number is used to specify the height of
the zoomed area. If the circles look a bit crude you can smooth them by
REGENerating the drawing. To draw the lines use **OSNAP** running mode
TANgent, and since you will be using **INTERSEC** shortly it can be included
in the command.

Command: **REGEN**

Command: **OSNAP** or pick Settings/Object Snap...

Object snap modes: **TANGENT,INTERSEC** or pick from dialogue
    box.

This selection of two modes means that AutoCAD will look for either a tangent
point or an intersection point every time a point is picked.

Command: **LINE**

From point: (Pick the small outer circle near point H)

To point: (Pick the largest circle near point J)

To point: <**ENTER**>

Note that the Object Snap mode overrides the ORTHO mode. Also note that
the line did not appear until the second point was picked. This was because
AutoCAD had to calculate the tangent point on the first circle and this was
dependent on the second point of the line. Now repeat this process for points
K and M.

Command: <**ENTER**>

LINE From point: (Pick the small outer circle near point K)

To point: (Pick the largest circle near point M)

To point: <**ENTER**>

Trimming entities

The two outer circles must now be trimmed back to their intersection points
with the tangents (Figure 7.7). The **TRIM** command works like EXTEND
and can be found in the Modify menu. There is also a Trim button in the
toolbox. With this command, you are first prompted for the boundary lines or
arcs, etc, to define the trimming edges. Then you specify the entities to trim.

Command: **TRIM**

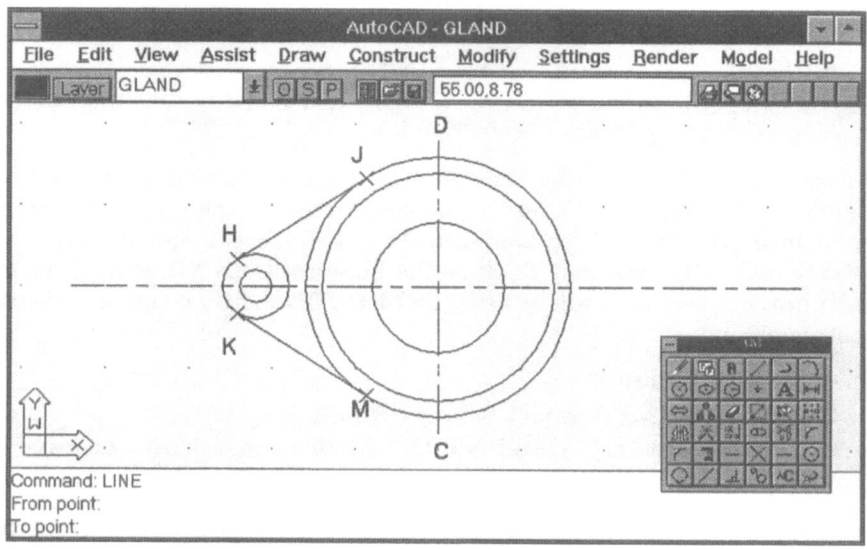

Figure 7.6 Tangential lines

Select cutting edge(s)...
Select objects: (Pick the line HJ near its mid point)
Select objects: (Pick the line KM near its mid point)

The two lines should now appear ghosted. If anything else has been selected
by mistake type **Remove** and pick the unwanted objects. When the selection
of the boundaries is completed press **<ENTER>** to proceed with trimming
the circles.

Select objects: **<ENTER>**
Select objects to trim: (Pick the small outer circle near N)
Select objects to trim: (Pick the largest circle near P)
Select objects to trim: **<ENTER>**

Before going any further you must reset the running **OSNAP** or object snap
mode to **None**. It is easy to forget about the object snap mode and that could
lead to undesirable results when picking points later.

Command: **OSNAP**
Object snap modes: **NONE**

Now to complete this view you can mirror the half-flange about the center-
line, CD. Pick **Construct/Mirror** from the menu bar.

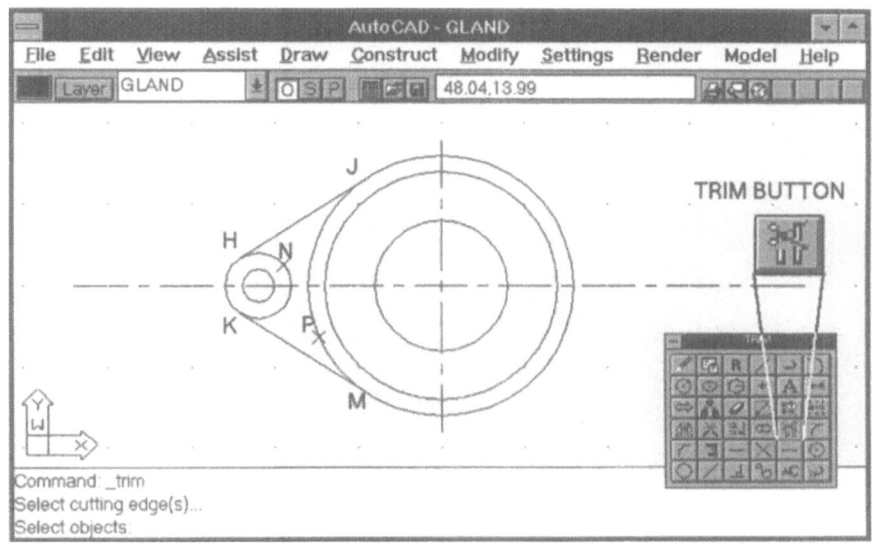

Figure 7.7 Trimming the flange

Command: **MIRROR**
Select objects: **Window**
First corner: **21,12**
Other corner: **37,29**
4 found.
Select objects: **<ENTER>**
First point of mirror line: **35,11**   (C)
Second point: (Pick a point vertically above C using ORTHO)
Delete old objects? **<N>**: **<ENTER>**

The large circle can be trimmed on the other side as before.

Command: **TRIM**
Select cutting edge(s)...
Select objects: **44,24**
1 found.
Select objects: **44,16**
1 found.
Select objects: **<ENTER>**
Select objects to trim: **43,19**
Select objects to trim: **<ENTER>**

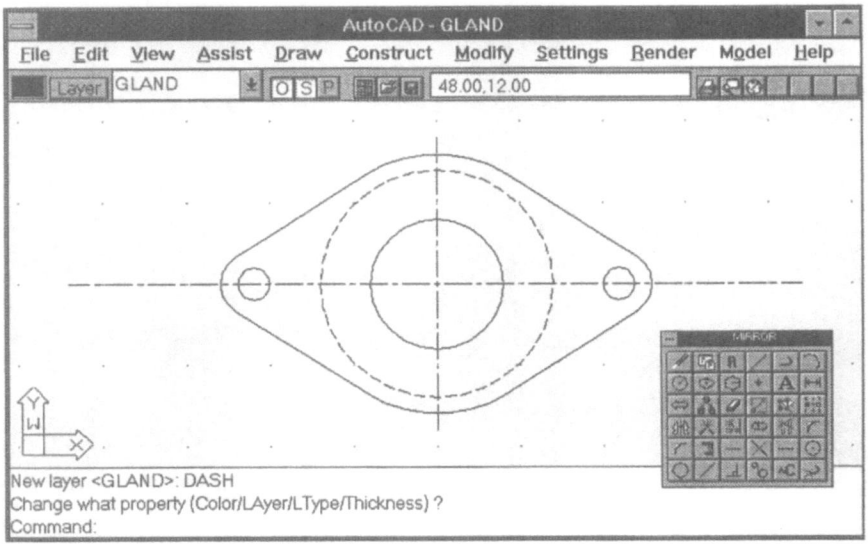

Figure 7.8 Plan view of gland

For the plan view to be complete the 14cm diameter circle should be drawn in dashed linetype (Figure 7.8). To do this change it to the **DASH** layer.

Command: **CHPROP**
Select objects: **42,21**
1 selected, 1 found.
Select objects: **<ENTER>**
Change what property (Color/LAyer/LType/
  Thickness)? **LA**
New layer <GLAND>: **DASH**
Change what property (Color/LAyer/LType/
  Thickness)? **<ENTER>**

Dimensioning

As the gland has been drawn to full scale, all the correct length information is already stored in the drawing. To extract this information and display it in the conventional way with dimension lines, etc, you will have to enter AutoCAD's "DIM" program. This is a sub-program of AutoCAD which is used to produce all the dimension lines semi-automatically and interactively. All the types of dimensioning normally found on engineering and architectural drawings are catered for, and as with the rest of AutoCAD you have complete control over how it is drawn.

In this section you will add horizontal and vertical dimensions, a diameter and radius and add center markings for the flange bolt holes. Before actually drawing any dimensions you should change layers and choose some settings for their display. The dimension text will use the font of the current text style. In the Gland drawing the font, *romans.shx* is used. Other fonts may give slightly different arrangements. For compatability set the text style to simplex. Pick **Draw/Text** and then **Set Style....** The pick **Roman Simplex** from the list followed by **OK**. Now press <**ENTER**> six times to accept the defaults and get back to the command prompt.

    Command: **LAYER**
    ?/Make/Set/...: **S**
    New current layer <GLAND>: **DIMENSIONS**
    ?/Make/Set/...: <**ENTER**>
    Command: **ZOOM**
    All/.../Window/<Scale(X/XP)>: **A**

When you enter the DIM: environment only commands that help with dimensioning are allowed. All toggles and object snapping are available but many of the usual AutoCAD drawing and editing commands are not. If things go wrong, ^C will always cancel the command and return you to the "Dim:" prompt. To execute commands other than dimensioning, you will have to exit from the "Dim:" prompt by typing **EXIT**. Picking commands from the pull-down menus will also exit the DIM: environment.

Once in DIM: you need to modify a number of parameters which control the sizes of the arrows, text etc. The default settings work well if you use the default drawing limits. However, the Gland limits are (65,45) which is much smaller than the default. You can experiment with dimension text heights and arrow sizes to get sensible results. To find out the current settings of all the dimension control variables type **DIM** followed by **STATUS**.

    Command: **DIM**
    Dim: **STATUS**

This gives a few pages of text information on the settings of over 40 system variables (Figure 7.9). To see the second page press <**ENTER**>. This is quite an eyeful of similar looking items. Don't despair as most of these are already set to the correct values. Their meanings will be explained in due course. The most important variables are the ones that control the text size and the arrow size. The default text size is given by the variable DIMTXT which has a value of 3.0000 units. The arrow size is also 3.0000 units, from the DIMASZ

DIMALT	Off	Alternate units selected
DIMALTD	2	Alternate unit decimal places
DIMALTF	0.0400	Alternate unit scale factor
DIMAPOST	"	Suffix for alternate text
DIMASO	On	Create associative dimensions
DIMASZ	3.0000	Arrow size
DIMBLK		Arrow block name
DIMBLK1		First arrow block name
DIMBLK2		Second arrow block name
DIMCEN	−3.0000	Center mark size
DIMCLRD	BYBLOCK	Dimension line color
DIMCLRE	BYBLOCK	Extension line and leader color
DIMCLRT	BYBLOCK	Dimension text color
DIMDLE	0.0000	Dimension line extension
DIMDLI	10.000	Dimension line increment for continuation
DIMEXE	1.5000	Extension above dimension line
DIMEXO	2.5000	Extension line origin offset
DIMGAP	0.0900	Gap from dimension line to text
DIMLFAC	1.0000	Linear unit scale factor

Press RETURN to continue:

Figure 7.9 Status of dimension variables

variable. As both of these are much too large for the current drawing they should be changed. Rather than having to change all the size variables for each drawing, AutoCAD provides an overall scale factor, DIMSCALE, which can be used to increase or decrease the sizes by its value. To get a reasonable display set DIMSCALE to 0.4. The variable, DIMASO, should be ON. If yours is off type **dimaso** and **ON**. This makes the dimension lines associative and allows them to be edited more easily. Furthermore, if one of the extension lines of an associative dimension is edited by the STRETCH command then the dimension text will automatically change to the new correct length.

Dimension style

Even though the default settings of all the dimension variables will give reasonable looking results there will be times when you will want to make alterations. Once you have found your favorite combination it can be saved as a "Dimension Style". Indeed, using this feature you can gain access to all of the variables in a more user friendly fashion.

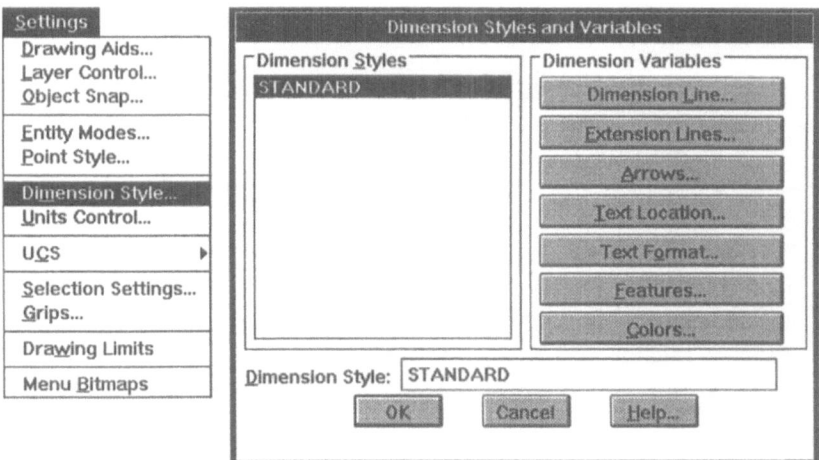

Figure 7.10 Dimension style

You can control the current display settings for the dimension lines and text by picking **Settings** from the menu bar followed by **Dimension Style** (Figure 7.10). All drawings have a default Dimension Style called "*UN-NAMED". It is recommended that you develop your own dimension style definition and save it. Move to the Dimension Style input field, just above the OK, Cancel and Help buttons, and input **STANDARD**.

The Dimension Styles and Variables dialogue box gives access to the 44 or so variables' settings. Most of the important ones can be found by picking the **Features** button. The resulting dialogue box should be filled in as shown in Figure 7.11. The most important number here is the feature scaling value of **0.4** sets the DIMSCALE variable mentioned above. This scaling is applied to all other dimension variables.

Moving down the left side of Figure 7.11, the text gap of **1.5** is the distance between the dimension text and the broken end of the dimension line. The baseline increment is used when generating multiple dimensions from one base point. It is the distance between one dimension line and the next. The actual arrow size will be the product of the feature scaling and the value in the arrow size box (0.4×3.0=1.2 units).

The right-hand side of the Features dialogue box concerns the extension lines and the text. The extension lines are the lines extending from the object to the arrows, eg the lines perpendicular to the arrows in Figure 7.13. The "extension above line" specifies how far above the arrow the extension line will go. The feature offset is the gap between the end of the extension line and the point picked. The center mark is a cross drawn at the center of circles

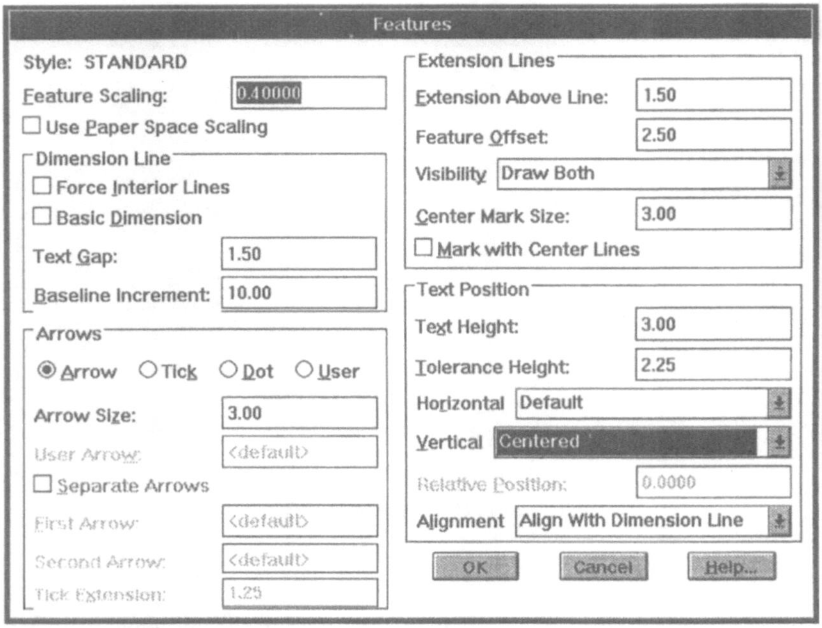

Figure 7.11 Dimension features

and arcs. The actual text height will be the product of the 3.0 and the feature scaling factor of 0.4. The tolerance height is the height of the text associated with ±range for dimension tolerances. The default horizontal text position is to have both arrows and text inside the extension lines as long as there is enough room. Otherwise the text and arrows will be placed outside. The vertical position can be either above the dimension line, centered in a break in the line or positioned relatively. To pick **Centered** use the pull down icon at the end of the input field. Then pick the keyword. Finally, choose the alignment to be parallel to the dimension line.

Warning! Dimension text can be aligned when inside the extension lines and horizontal when outside. To do this set the variable DIMTIH to On (Text Inside Horizontal? No thanks!) and DIMTOH to On. The alignment options in some versions of the menu dialogue boxes do the opposite of what they are supposed to do. This is a bug in the menu.

When your Features dialogue box matches Figure 7.11 pick **OK** to get back to the styles dialogue box (Figure 7.10). The other buttons above the Features one give a bit more detail on the various options but not much extra. The Colors... button can be quite useful. This allows you to set different colors for the extension lines, the arrows and the text. By doing this you can plot dimensions using thin lines and normal thickness for the text.

The above procedure was rather involved. If you are in doubt as to what values your dimension variables should have you can set the more important ones as follows. You can type **DIMSCALE** at either the "Dim:" or "Command:" prompts.

Dim: **DIMSCALE**
Current value <1.0000> New value: **0.4**
Dim: **DIMTXT**   (Text size)
Current value <default> New value: **3**
Dim: **DIMASZ**   (Arrow size)
Current value <default> New value: **3**
Dim: **DIMASO**
Current value <default> New value: **ON**
Dim: **DIMTAD**   (Put text above dimension line?)
Current value <default> New value: **OFF**
Dim: **DIMTIH**   (Text inside horizontal?)
Current value <default> New value: **OFF**
Dim: **DIMTOH**   (Text outside horizontal?)
Current value <default> New value: **OFF**

Having set the scene, let's see how it works out. The actual dimensioning commands can be found in the **Draw** pull-down menu (Figure 7.12). The **Dimensions** sub-menu has five items. The most important of these is the **Linear** sub-menu. Horizontal and vertical dimensioning will be used on the Gland. Rotated and aligned will be used later on a polygon object. Baseline and continue are useful for generating a line of dimensions or a set of running dimensions. The **Radial** sub-menu gives commands for drawing either a diameter or radius dimension. A center mark can also be added. The **Ordinate** menu is useful for dimensioning points relative to some datum. **Angular** allows the angle between objects to be dimensioned. Finally, the **Leader** command draws an arrow with a leader line to an annotation.

To dimension the flange of the gland pick **Draw/Dimensions** and **Linear**. Then pick **Horizontal**. You are prompted for the "First extension line origin". Pick the leftmost point of the flange (point a on Figure 7.13). You could use the object snap INTERSEC for this. Pick the furthest right point for the second extension line origin. After selecting the two points to be dimensioned you are asked where you want the actual dimension line to be drawn. Give any point whose y coordinate is 38 units.

Command: _dim1
Dim: **horizontal**
First extension line origin or RETURN to select: **INTERSEC**

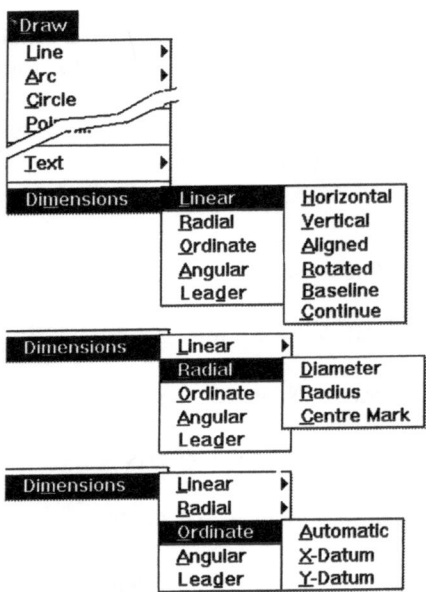

Figure 7.12 Dimension menus

of **22,20**   (Leftmost point on flange)

Second extension line origin: **48,20**   (Other end of flange)

Dimension line location (Text/Angle): **50,38**

Dimension text <26.0000>: <**ENTER**>

Finally, you are asked for the text to put on the dimension line. AutoCAD calculates the horizontal distance between the two extension line origins and offers that as the default. Four places of decimals are given since that is the current UNITS setting. Pressing <**ENTER**> accepts the default and the dimension is included as shown in Figure 7.13. You can override the default by simply typing whatever text you wish. The text is aligned with the dimension line which is broken. The "Text/Angle" option in the "Dimension line location" prompt allows you to check the text before inputting a location and override the alignment by giving an angle for the text.

If your dimension is not correct, type **undo** to erase it and try again. If the text is too large or too small, check the values of DIMSCALE, DIMTXT and DIMASZ. Type **DIM** followed by **STATUS** to do this.

The "horizontal" dimensioning command produces the horizontal distance between any two points. The distance is calculated from the X coordinates of the two extension line origins. The "vertical" operates in a similar fashion

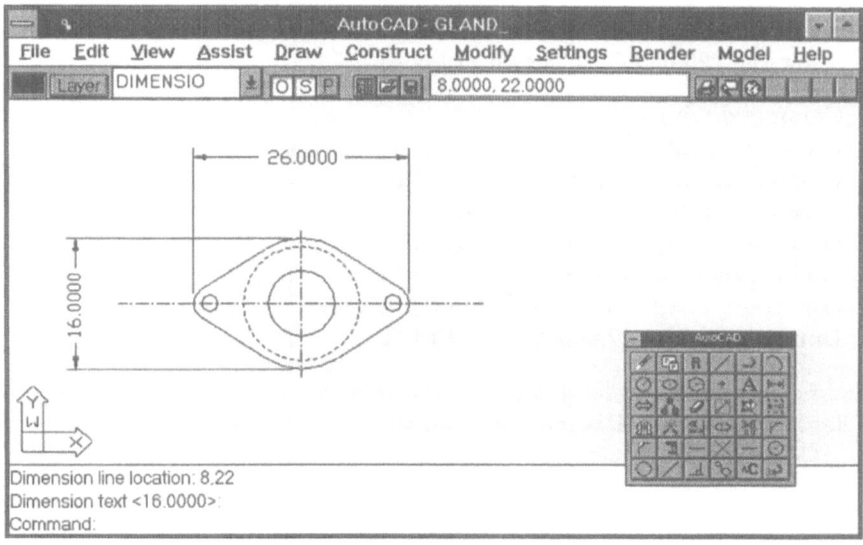

Figure 7.13 Overall dimensions

for vertical distances calculated from the Y coordinates. Pick **vertical** from the screen menu and give the quadrant points at c and d as the extension line origins. The dimension line should be located at (8,22) or nearby. Note that as you pick commands from the **Draw/Dimensions** pull-down menu AutoCAD starts the process by issuing "dim1" at the command prompt. This puts AutoCAD in the "Dim:" environment for one operation only. At the end of the sequence you are returned to the "Command:" prompt. If you are typing the commands use DIM rather than DIM1 to remain in the environment for the next operation.

> Command: _dim1
> Dim: **vertical**
> First extension line origin or RETURN to select: **quad**
> of (Pick point c near 35,28)
> Second extension line origin: **quad**
> of (Pick point d near 35,12)
> Dimension line location: **8,22**
> Dimension text <16.0000>: **<ENTER>**

To add the distance between the flange bolt holes and the radius of the smaller flange arc use **horizontal** once more followed by **continue.** Pick the centers of the two bolt holes as the extension line origins and position the new dimension line below the "26.0000", as shown in Figure 7.14.

> Command: _dim1
> Dim: **horizontal**
> First extension line origin or RETURN to select: **cen**
> of (Pick bolt hole circle at 24,20)
> Second extension line origin: **cen**
> of (Pick right hand hole at 46,20)
> Dimension line location(Text/Angle): **45,35**
> Dimension text <22.0000>: <**ENTER**>

Now to continue the dimension line to the right for the distance to the end of the flange pick **Draw/Dimensions** and then **Linear/Continue**.

> Second extension line origin: **48,20**
> Dimension text <2.0000>: <**ENTER**>

This should give the new horizontal dimension lines shown in Figure 7.14. The "2.0000" is too long to fit between the extension lines so AutoCAD automatically places it outside. Continue can also be used for vertical dimensions.

You don't have to pick the two extension line origins if the object being dimensioned is made from just one entity. All you have to do is press <ENTER> and select the object. As an alternative to using the mouse to pick all the menus you could use the hot keys. Press **ALT** followed by D for Draw, M for Dimensions, L for Linear and H for Horizontal. Pressed in quick succession this is not too bad ie ALT D M L H. If you press the wrong letter and end up on the wrong sub-menu press the ESC key to go back a step.

Now, rather than giving the extension line origins press <**ENTER**> to select the object and pick the circle.

> Command: _dim1
> Dim: **horizontal**
> First extension line origin or RETURN to select: <**ENTER**>
> Select line, arc, or circle: **30,25** (Point on dashed circle)
> Dimension line location (Text/Angle): **30,32**
> Dimension text <14.0000>: <**ENTER**>

To demonstrate the use of the "radius" and "diameter" commands, the right-hand bolthole and the central circle will now be dimensioned (Figure 7.15).

Pick **Draw/Dimensions** and **Radial/Radius** from the pull-down menu. Then pick the small circle at the right hand side of the gland. As the point on the circle that you pick determines where the dimension line will be drawn, you should take some care to pick the circle in its lower left quadrant (near the point 23.3,19.3). If snap and/or ortho are currently on toggle them off using

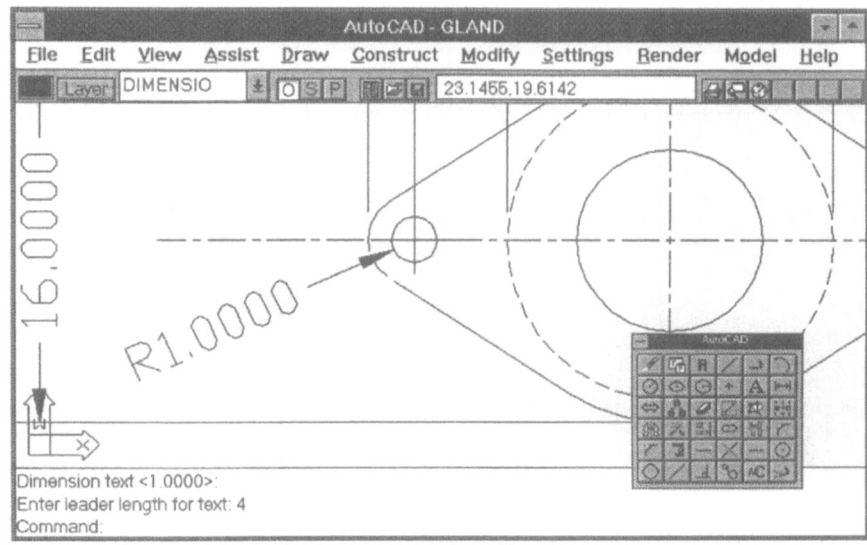

Figure 7.14 More horizontal dimensions

the **CTRL** key with **B** and then with **CTRL O** or use the S and O buttons
on the tool bar. In the sequence given below a transparent zoom command
is used to get a closer look. When the dimension command resumes, pick the
circle and accept the default dimension text and as this does not fit inside the
circle you will be prompted for a leader length.

> Command: dim1
> Dim: **Radius**
> Select arc or circle: **'ZOOM** or pick View/Zoom/Window
> >>Center/.../Window/<Scale(X/XP)>: **W**
> >>First corner: **7,10**
> >>Other corner: **29,26**
> Resuming DIM command:
> Select arc or circle: ^B<Snap off> ^O<Ortho off> (Pick point 23.3,19.3)
> Dimension text <1.0000>: **<ENTER>**
> Enter leader length for text: **4**
> Dim:

The apostrophe or quote before the zoom executes the transparent version of
the command which means that it can be run in the middle of other commands
including DIM commands. This command is located in the View pull-down
menu. The options list for this version of zoom is limited and does not include

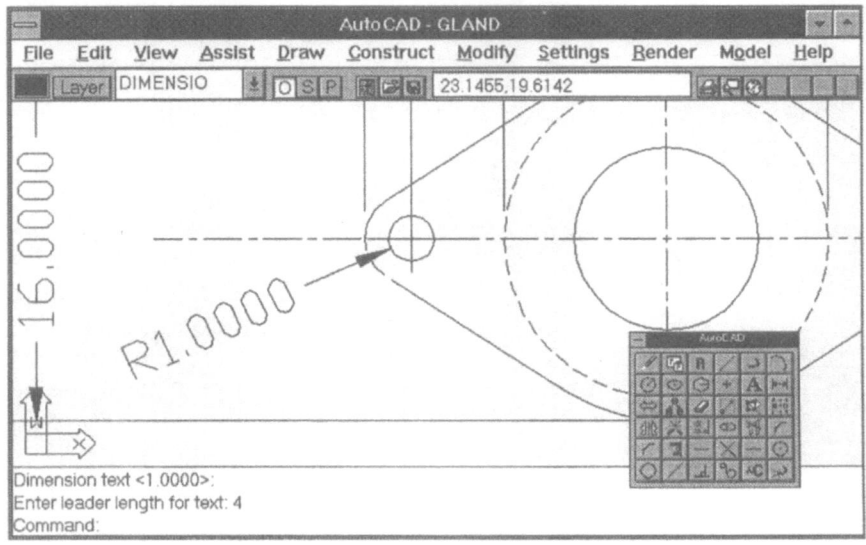

Figure 7.15 Radius and 'ZOOM

zoom, all for example. Otherwise it works much the same as the usual zoom command.

The leader is the arrow and line leading from the object to the text. If you accept the default text offered by the radius command, the letter "R" is automatically prefixed to the number. This command also draws a cross at the center of the circle. The size of the cross is determined by the value of the DIMCEN and DIMSCALE variables. These were set as the Feature scaling and Center mark size in Figure 7.11.

Before including the diameter for the internal circle, use the zoom to return to the previous magnification. Pick **Diameter** from the **Radial** submenu or type "DIAM" at the Dim: prompt. Then pick the inside circle near the point (38,17). Accept the text offered by AutoCAD and give a leader length of 6.

Command: **ZOOM**
All/.../Previous/.../<Scale(X/XP)>: **P**
Command: _dim1
Dim: **Diameter**
Select arc or circle: **38,17**
Dimension text <8.0000>: **<ENTER>**
Enter leader length for text: **6**

This time accepting the default text causes the diameter mark ϕ to appear in front of the text. This is the standard symbol for indicating a diameter dimension. Again the point picked on the circle controls the position of the dimensions.

To draw the center mark for the right-hand bolt hole pick **Center Mark** from the **Radial** sub-menu and pick the circle near the point (46,19).

> Command: _dim1
> Dim: **CENTER**
> Select arc or circle: **46,19**

As before, the size of this mark is dictated by the value of dimscale multiplied by dimcen. One variation is possible by specifying a negative value for the variable, dimcen. The negative value causes center-lines to be drawn which intersect the circle itself. In that case the size depends on the diameter of the circle. This can be set by picking "Mark with Center Lines" in the Features dialogue box (Figure 7.11). A dimcen value of zero suppresses the drawing of center marks with the radius and diameter commands.

To conclude the gland drawing (Figure 7.16) put a leader line to indicate that the sloping line is tangential to the arcs. The **leader** command allows you to place pointers with text on the drawing. Once you have picked **leader** from the **Draw/Dimensions** menu you will be prompted for the leader start point. This is where the point of the arrow will be drawn. Use the object snap **intersec** to locate the intersection of the arc and line near the point (47,18). The prompt changes to the familiar "To point:" request similar to the LINE command. Pick the point (51,16). You can make the leader line consist of as many line segments as necessary by picking points. When the line is completed press <**ENTER**> to exit the "To point:" prompt. Then type the desired text, which is the word "tangential".

> Command: _dim1
> Dim: **LEADER**
> Leader start: **INTERSEC** of **47,18** (or pick a point near to this)
> To point: **51,16**
> To point: <**ENTER**>
> Dimension text <8.0000>: **tangential**

Even though you gave only two points on the leader line, a final horizontal line segment is drawn automatically. The previous dimension text is offered as the default but in this case it is not suitable. The only real problem that can occur with drawing leader lines is if the distance between the start point and the first point is not long enough to draw the arrow. If this happens the command continues as normal but no arrow is drawn.

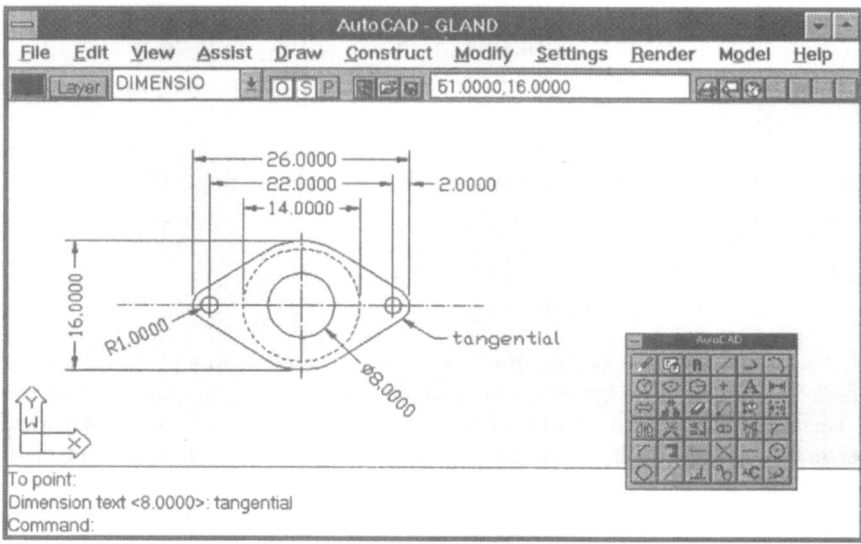

Figure 7.16 The dimensioned gland

Editing Dimensions

Now that all the dimension lines and text have been added you can explore some of the editing facilities for associative dimensions. One of the undesirable features of the dimensions in Figure 7.16 is that the four places of decimals are not really necessary. The dimension text of "2.0000" to the right can be modified to read "R2.00" and the radial and diameter text can be made horizontal.

To set the radius and diameter text to be horizontal set the dimension system variable, **DIMTOH** (DIMension Text Outside is Horizontal) to be 1 or **On**.

Command: **DIMTOH**
New value for DIMTOH <0>: **1**

Warning! There is an option in the Dimension Style, Features dialogue box which should do this. Pick **Settings/Dimension Style...** followed by picking **Features**. Then picking the **Alignment** pull-down arrow, towards the bottom right of the screen, you should see the option "Aligned When Inside Only". This should have the same effect as changing DIMTOH as above. With some versions of the menu interface it does the opposite! This is a little bug with AutoCAD.

The precision of dimensions is controlled by the drawing UNITS setting. At the start of the Gland drawing the units were set to decimal with four places after the decimal point. Pick **Settings/Units Control...** from the menu bar. Then pick the **Precision** as **0.00** as shown in Figure 7.17. You may have to use the slider bar to move up to the correct value. Make sure that the angular precision is 0. Then pick **OK**. This setting does not automatically change the dimensions already drawn but does effect all new dimension entities. To apply it to existing dimensions you must use the Dimension, Update command.

This command and other editing commands are found by picking **Modify** and **Edit Dimensions**. The resulting sub-menu shown in Figure 7.17. Now pick **Update Dimension**. You are then asked to select the objects. Use a window to surround everything.

Command: _dim1
Dim: **Update**
Select objects: **Window**
First corner: **0,0**
Other corner: **65,45**
30 found
Select objects: **<ENTER>**

Only dimension entities will be modified. Furthermore, only associative dimensions that were created using the default text will be modified. To change the updated text of "2.00" to "R2.00" pick **Modify/Edit Dimensions** followed by **Dimension Text/Change Text** as shown in Figure 7.17. You will be prompted to enter the new text first and then asked to pick the entity to be modified.

Command: _dim1
Dim: **NEWTEXT**
Enter new dimension text: **R2.00**
Select objects: **53,36** (Point near the text "2.00")
1 found
Select objects: **<ENTER>**

The other text modification functions allow you to **Move Text** and **Rotate Text**. Both are self explanatory and useful when the drawing becomes congested. The **Hometext** option returns dimension text to the default position, based on the dimension style settings. It can also be used to reverse the 2.00 to R2.00 edit above. Hometext also comes in handy after a dimension line has been STRETCHed.

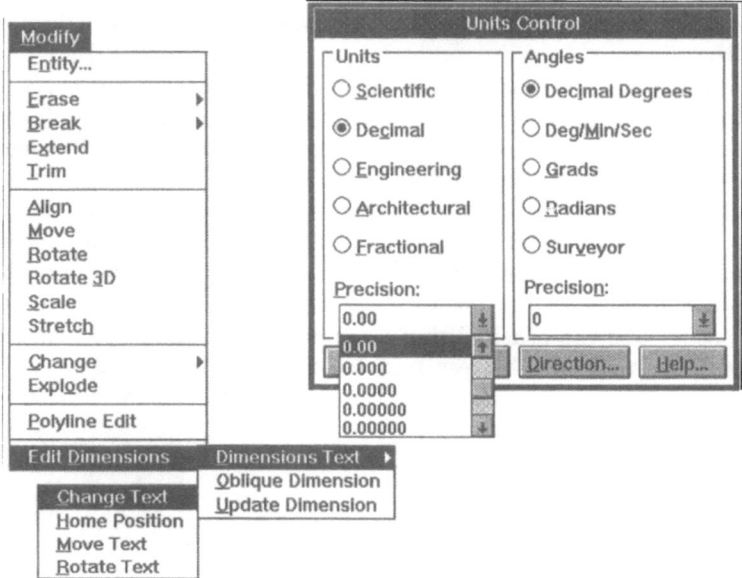

Figure 7.17 Editing dimensions

The result of this alteration is shown in Figure 7.18. The other alterations shown in the diagram are the result of a combination of changes in the dimension variables, the drawing units and the UPDATE command. These will now be described in detail. This is a good time to save your drawing and take a break.

Standard dimension variable settings

Most drawing offices have a house style for dimensioning drawings. By setting all the parameters and creating a named dimension style you can quickly access the house style. In the United Kingdom, British Standard number 308 gives guidance on how dimensions should be laid out. To comply with BS308 the dimension variables should be set as given in Table 7.2. This gives a commentary on the meanings of the values.

The first four variables are concerned with settings for alternate units to be included in the drawing. As the international system of units is in millimeters a second system is rarely required. However, some disciplines are in a transition from imperial units to metric and so may wish to include both measurements.

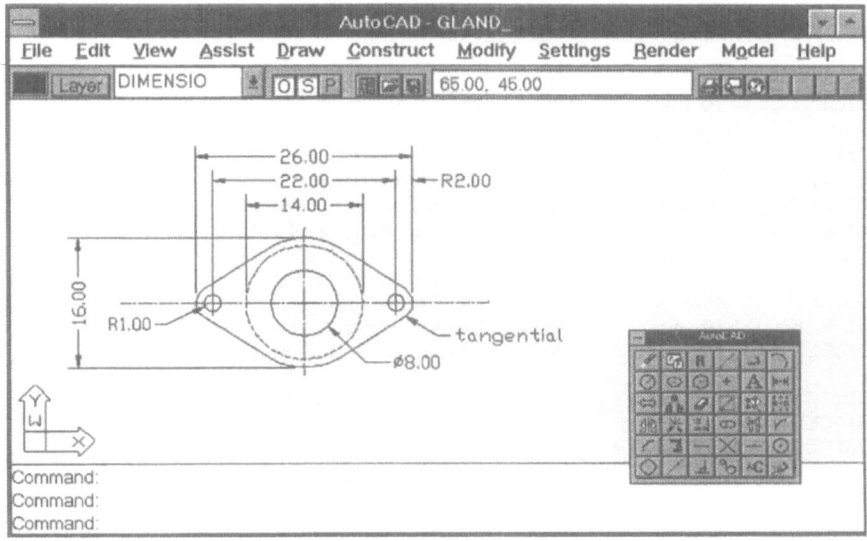

Figure 7.18 The updated dimensions

The values given in Table 7.2 assume that if alternate units are to be used, then the main units are millimeters and the secondary units are inches.

Make sure that all your dimension variables are set to the values given in Table 7.2. The quickest way of doing this is by typing **STATUS** at the Dim: prompt. Any variables that differ can be changed by typing their name and then giving the appropriate value.

Having selected all the appropriate dimension variable settings, the current status can be saved in a named "dimension style". The named style can be selected for use later by using the DIM: RESTORE command or by picking **Settings/Dimension Style** from the menu bar. If you do this in your prototype drawing then the style will be available each time you start a new drawing.

> Dim: **SAVE**
> ?/Name for new dimension style: **BS308**
> Dim: **EXIT**

Drawing a pentagon

To illustrate the remaining dimension commands and a couple of new drawing commands, you will now create a pentagon and find out the internal angle

Table 7.2 Dimension variable settings for BS308

Variable name	BS308 setting	Explanation
DIMALT	Off	Alternate dimensions not used.
DIMALTD	2	Decimal places for alternate units.
DIMALTF	0.0394	Factor for alternate units (number of inches in 1mm).
DIMAPOST	"	Will suffix the inches symbol after alternate units.
DIMASO	ON	Enables associative dimensioning.
DIMASZ	3.00	Size of arrow to be used to terminate dimension lines.
DIMBLK	None	Name of block to be used in place of the arrows.
DIMBLK1	None	Block to be used at first extension line instead of arrow.
DIMBLK2	None	Block to be used at second extension line. DIMBLKx only used with DIMSAH On.
DIMCEN	−3.00	Size of center mark. Negative value gives center-lines.
DIMCLRD	BYBLOCK	Dimension line color. Set to color for thin lines.
DIMCLRE	BYBLOCK	Extension line color. Same as DIMCLRD.
DIMCLRT	BYBLOCK	Dimension text color. Color for normal thickness pen.
DIMDLE	1.25	Extend dimension line. For use with DIMTSZ only.
DIMDLI	10.00	Distance between dimension lines using BASELINE.
DIMEXE	1.50	Length of extension line beyond arrows. AutoCAD default is too large.
DIMEXO	2.50	Extension line offset.
DIMGAP	1.50	Gap either side of text if DIMTAD is off and DIMTVP 0.
DIMLFAC	1.00	1 drawing unit = 1 dimension unit.
DIMLIM	OFF	Limit dimensions not required. See DIMTOL.
DIMPOST		Unit suffix for dimensions. None required.
DIMRND	0.00	Rounding off value for dimesnions. Not required.
DIMSAH	Off	Disables DIMBLK1 and DIMBLK2.
DIMSCALE	0.40	Overall scaling factor for size of arrows and text. Depends on drawing size. Text should be 3mm on plots.
DIMSE1	Off	Disables suppression of first extension line. Would turn on for running dimensions.
DIMSE2	Off	Disable supression of second extension line.
DIMSHO	On	Turn on dynamic display of dimension value.
DIMSOXD	Off	Disable supression of dimension line drawn outside the extension lines.
DIMSTYLE	EXPRESS	Name of current dimension style.
DIMTAD	On	Draw dimension text above dimension line. See polygon.
DIMTFAC	0.75	Scale factor for tolerance text height.
DIMTIH	Off	Makes text inside extension line parallel to dim line.
DIMTIX	Off	If this is on, it forces text inside extension lines.
DIMTM	0.00	Minus value for tolerance dimensions. See DIMTOL.
DIMTOFL	Off	If this is on, it forces text outside extension lines.
DIMTOH	Off	Makes text outside extension lines parallel to dim line. Text will be drawn horizontally if On.
DIMTOL	Off	If on, this enables tolerance dimensions.
DIMTP	0.00	Plus value for tolerance dimensions. See DIMTOL.
DIMTSZ	0.00	Tick size. If non-zero, ticks are drawn instead of arrows.
DIMTVP	0.00	Vertical position of text. Effective if DIMTAD is OFF.
DIMTXT	3.00	Size of dimension text.
DIMZIN	1	Include zero feet and inches units, 4 causes all leading zeros of decimal units to be supressed. See Help, DIMZIN for all settings.

between two adjacent sides. To save time in setting up a new drawing environment just change to the POLYGON layer and freeze the others. This will allow you to use all the current dimension variable settings. Some of these will be altered temporarily to create tolerant dimensions.

> Command: **LAYER**
> ?/Make/Set/...:S
> New current layer <DIMENSIONS>: **POLYGON**
> ?/Make/.../Freeze/Thaw: F
> Layer name(s) to Freeze: *
> Cannot freeze layer POLYGON. It is the CURRENT layer.
> ?/Make/Set/...: <**ENTER**>

This method of freezing all the layers but the current one is much faster than using the Layer Control dialogue box.

The **Polygon** menu can be found in the **Draw** pull-down menu. There are three choices for creating a polygon. It can be inscribed in a circle (the vertices touch the circle) of a given radius. It can circumscribe a circle (sides are tangential to the circle). Here we will specify the length on one edge. Pick **Edge** from the polygon menu.

When you select this command you are first prompted for the number of sides. Use 5 sides for a pentagon. You can then either specify a circle to be inscribed or circumscribed by the polygon or you can give the position and length of one side.

> Command: **POLYGON**
> Number of sides: **5**
> Edge/<Center of polygon>: **EDGE**
> First endpoint of edge: **20,10**   (A)
> Second endpoint of edge: **@20,0**   (B)

This actually draws a closed polyline, calculating the vertices from the geometrical properties of equilateral polygons. It can be edited in the same way as any other closed polyline.

The CHAMFER edit command can be used to cut off the top corner. This command is similar to the FILLET command but draws a straight line between the chamfer points. For this command you give the length by which each of a pair of lines is to be trimmed back. If a polyline is to be chamfered then you have the further option of trimming all the corners. For example, to chamfer the corners of the pentagon by trimming 3 units from each end of the line segments you would get the shape given in Figure 7.19. You first have to give the sizes of the chamfer and then the polyline to be edited. Pick **Construct/Chamfer.**

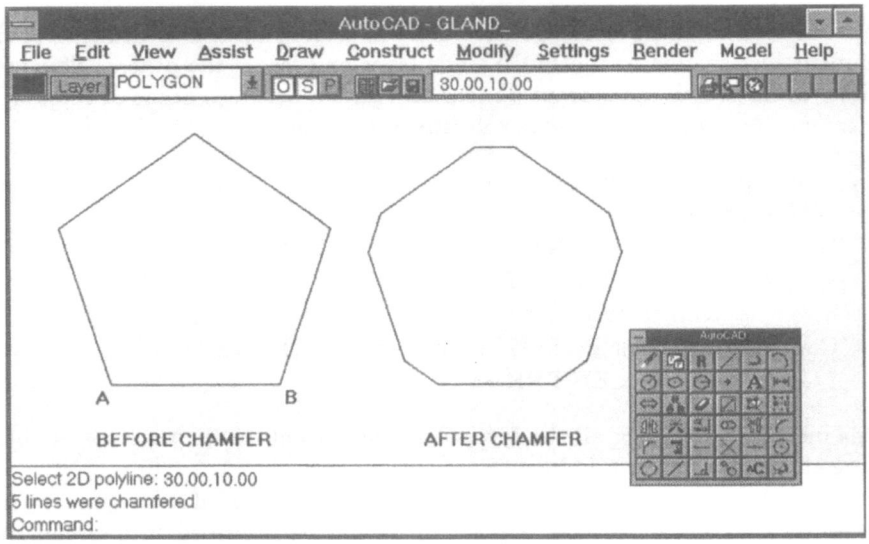

Figure 7.19 The chamfered pentagon

Command: **CHAMFER**
Polyline/Distance/<Select first line>: **DISTANCE**
Enter first chamfer distance <0.00>: **3**
Enter second chamfer distance <3.00>: **<ENTER>**
Command: **<ENTER>**   (To re-execute the command)
CHAMFER Polyline/Distance/<Select first line>: **Polyline**
Select 2D polyline: **30,10**

Chamfer can also be used with unequal distances and be applied to individual pairs of lines. If a polyline to be chamfered contains an arc then the arc will be deleted and replaced with a straight line.

Wrapping up dimensions

The "angular" option in the Dimension menu allows the dimensioning of angles between lines. To draw the angle between the longer lines AB and BC, pick **Draw/Dimensions** and **angular**. Then pick points on lines AB end BC. Indicate where the dimension arc is to be located and accept the default dimension text and text location.

Command: _dim1

Dim: **ANGULAR**
Select arc, circle,line: **30,10**   (Line AB)
Second line: **43,20**   (Line BC)
Dimension arc line location (Text/Angle): **31,10**
Dimension text <108>: <**ENTER**>
Enter text location (or RETURN): <**ENTER**>

Accepting the default text location causes the "108°" to be positioned in the middle of the arc. The dimension should look like that in Figure 7.20. Giving any other location response will put the text at that location.

With aligned dimensions the length is measured parallel to the line joining the two extension line origins. To find the new length of the line between C and D pick **Aligned** from the Linear sub-menu. Instead of picking the origin points press <**ENTER**> and then pick the line CD at the point (38,35). Put the dimension line at (42,39) and accept the default text.

Command: _dim1
Dim: **ALIGNED**
First extension line origin or RETURN to select: <**ENTER**>
Select line,arc, or circle: **38,35**
Dimension line location (Text/Angle): **42,39**
Dimension text <14.00>: <**ENTER**>

The dimension line is aligned with the line segment and gives the correct length. The original length was 20 from which 3 was taken from each end.

To dimension the chamfer at the point E use the **rotated** dimensions at an angle of 198 degrees. This angle is perpendicular to the line AE. Pick **Rotated** from the LINEAR sub-menu, then press <**ENTER**> and pick the short line at point E. Place the dimension line at (13,30) and accept the default text.

Command: _dim1
Dim: **ROTATED**
Dimension line angle <0>: **198**
First extension line origin or RETURN to select: <**ENTER**>
Select line,arc, or circle: **15.5,28.5**
Dimension line location (Text/Angle): **13,33**
Dimension text <2.85>: <**ENTER**>

Note that rotated dimensions with angle zero are the same as horizontal dimensions and angle = 90 gives vertical dimensions.

Finally to produce dimensions with a tolerance level built in change the Dimension style settings. Click the EXPRESS style. Then move to the Dimenstion Style name input field and change the name to EXPTOL. Then pick the **Text Format** button. This creates a new style based on EXPRESS but

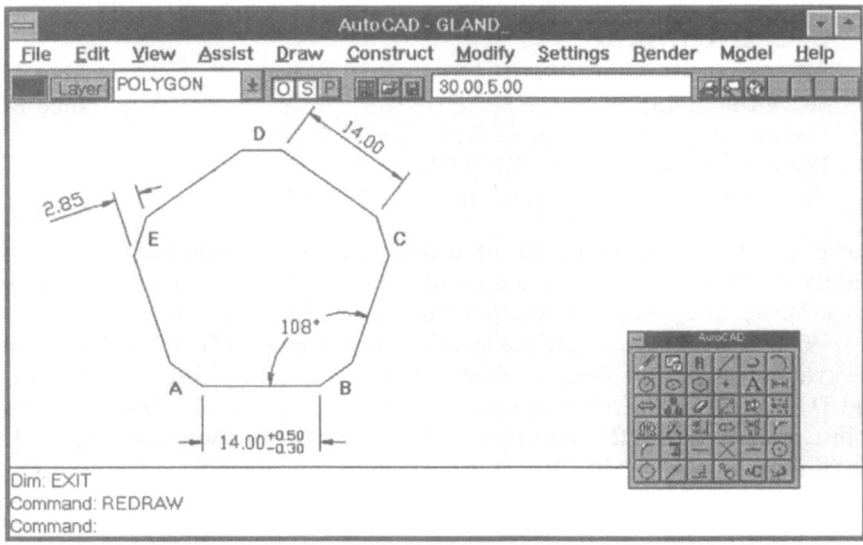

Figure 7.20 Final dimensions

including the following changes. The tolerance is controlled on the right of the Text Format pop-up. Pick **Variance** and give the upper value as **0.50** and the lower as **0.30** (Figure 7.21). Then pick **OK**. Pick **Text Location...** from the Dimension Style pop-up and set the horizontal text position to **Force Text Inside** as shown in Figure 7.21. Change the vertical text position to **Centered**. Pick **OK** twice and use **Linear/Horizontal** to dimension the line AB.

Note that if you do not make a new style for this last dimension all the dimensions will automatically be updated to tolerance ones. If you change the settings in a specific style then all dimensions drawing in that style will be updated.

Command: _dim1
Dim: **Horizontal**
First extension line origin or RETURN to select: **<ENTER>**
Select line,arc, or circle: **30,10**
Dimension line location: **30,3**
Dimension text <14.00>: **<ENTER>**

This should give the dimensions as shown in Figure 7.20. If you don't accept the default dimension text the tolerance values will not be drawn.

To finish this exercise, pick **File/Save** from the menu bar followed by **File/Exit AutoCAD**.

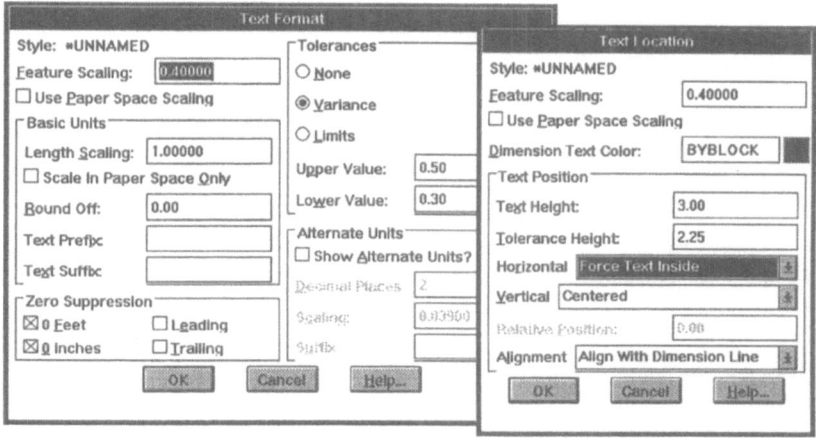

Figure 7.21 Tolerant dimensions

Summary

In this chapter you have encountered some advanced drawing and editing commands. Some of these, such as TRIM and CHAMFER allow you to dispense with having to draw preparatory construction lines. Others like POLYGON and DIVIDE draw multiple objects. By far the most important component covered in this exercise has been the dimensioning sub-system. AutoCAD changes when you are in Dim: and many new commands are made available while at the same time most of the drawing and editing commands are withdrawn.

Dimensions are calculated automatically from the current drawing units. It is important to choose suitable units and accuracy levels for sensible dimension values. The dimension environment can be tailored to your specific needs by setting up the relevant dimension variables and saving dimension styles.

You should now be able to:

Draw lines tangential to two circles.
Draw equilateral polylines.
Trim circles and chamfer polylines.
Use ZOOM center with a magnification.
Use the transparent zoom.
Add horizontal and vertical dimension lines.

Draw aligned, rotated, angular and tolerant dimensions.
Set up the dimension variables.
Edit existing dimensions.
Save a dimension style.

Chapter 8 ADDING DEPTH TO YOUR DRAWINGS WITH 3D CAD

General

Since the early days of AutoCAD, the programmers have been striving to produce full three-dimensional capability for the PC-CAD user. With Release 10, in 1988, they can justifiably claim to have done just that. Earlier versions of AutoCAD allowed only use of isometric projection techniques and a pseudo-3D known as 2.5D. This chapter will take a brief look at all three techniques to produce some simple drawings. These include an isometric cooker, an office building and the great pyramid of Giza (Figure 8.1).

The opening up of the Z axis brings new and exciting aspects to AutoCAD use. Things can be constructed on the computer screen at full scale and depth. Once the object has been drawn it can be viewed from above (plan), from the front and side (elevation) and in either isometric or perspective projection. You can "walk" around the AutoCAD image and even through it. These facilities are particularly useful for disciplines where it is necessary to have a full appreciation and visualisation of the design.

As a cautionary note, one should not get carried away with the novelty and hype associated with 3D CAD. Architects and engineers have successfully managed to develop the most complex of projects over hundreds of years using

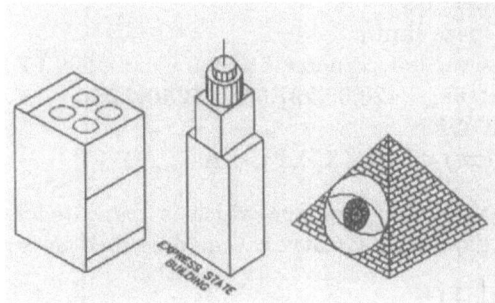

Figure 8.1 Pyramids and Towers

simple 2D drawings. Thus, for a lot of design projects the 2D representation is adequate. Any changeover to 3D CAD must justify the extra effort required. You should also be aware of what AutoCAD 3D can and can't do.

AutoCAD models solid objects as wire frame skeletons. As such you can see through the objects that are drawn. You can tell AutoCAD to HIDE the lines at the back of the object to give the impression of solidity.

All the commands you have used up to now also work in 3D, although some special commands are needed (eg 3DPOLY line). There is a whole new vocabulary of terms relating to 3D geometry and a set of completely new functions. Let the work commence.

Isometric projection

Isometric projection is still the standard method of conveying three- dimensional engineering information on a two-dimensional sheet of paper. To produce anything other than simple shapes in isometric projection requires considerable expertise in drafting techniques. It is not my purpose to introduce such drawing construction methods but I do wish to display the special features within AutoCAD for isometric projections. To demonstrate these features and the basics of isometric projection we will create a drawing of the cooker that was used in Chapter 6.

Start up AutoCAD, pick **File/New** to create a new drawing. Call it **COOKISO** with ACAD as the prototype. Set the LIMITS to (0,0) and (3250,2250) and the UNITS to Decimal. Set the GRID to 100 units and SNAP to 50 using the Drawing Aids dialogue box. This pop-up dialogue box is also used to set up the isometric axes (Figure 8.2.

> Command: **LIMITS**
> Reset Model space limits:
> ON/OFF/<Lower left corner> <0.00,0.00>: <**ENTER**>
> Upper right corner <420.00,297.00>: **3250,2250**
> Command: **ZOOM**
> All/.../Window/<Scale (X/XP)>: **A**

Setting units gives a lot of dialogue which is truncated below. You can, of course use **Settings/Units Control...** as described in earlier chapters.

> Command: **UNITS**
> Report formats:...
> Enter choice, 1 to 5 < >: **2**
> Number of digits to right of decimal point (0 to 8) <2>: **1**
> System of angle measure:...
> Enter choice, 1 to 5 < >: **1**

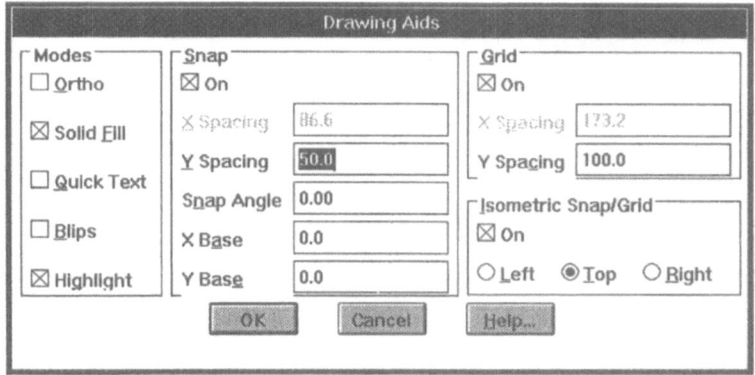

Figure 8.2 Setting Isometric Snap and Grid

Number of fractional places for display of angles (0 to 8) <4>: **2**
Enter direction for angle 0.00 <0.00>: **<ENTER>**
Do you want angles measured clockwise? <N> **<ENTER>**

These settings will be useful for the other two drawings in this chapter. To keep them safe, SAVE the drawing with the filename **EXPROTO**. This will be used as a prototype drawing later. It contains the correct limits and unit settings.

Command: **SAVE**

Pick the **Type it** button to use the command line.

Save current changes as <COOKISO>: **EXPROTO**

There is a subtle difference between typing SAVE and using the File menu. By typing the SAVE command, the current drawing name in the title bar, at the top of the window remains COOKISO. If you pick File/Save As the title would change to EXPROTO.

Now to set the Grid and Snap and to switch on the isometric projection pick **Settings/Drawing Aids...** from the menu bar. When the dialogue box (Figure 8.2) appears, click the Snap **On** with a value of **50** for X and Y. Similarly, set the Grid to **On** with **100** for the X and Y values. Then pick the **On** box for **Isometric Snap/Grid** and pick the circle button beside **Top** as shown.

As soon as Isometric is switched on the X spacings for snap and grid change to 86.6 and 172.2 and become ghosted. The values of the X spacing in isometric are restricted to twice the Y value multiplied by the sine of 60°. This

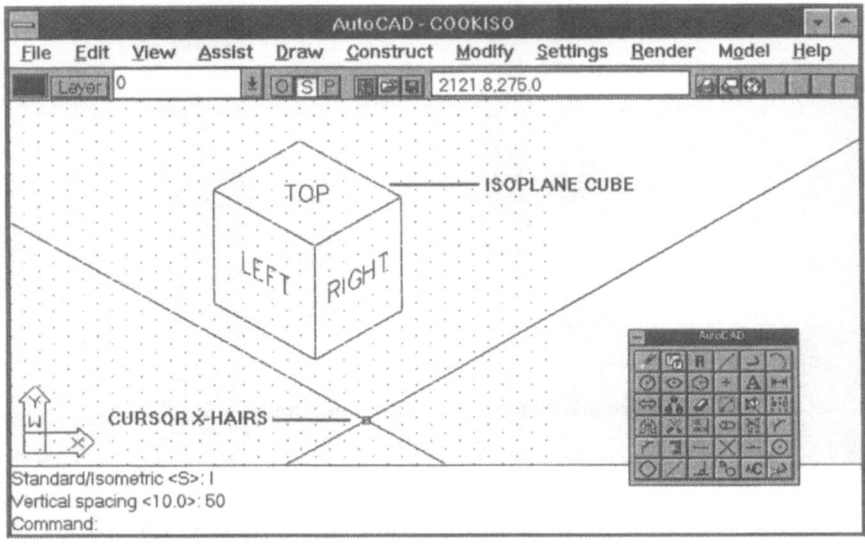

Figure 8.3 Isometric screen

is because the "horizontal" isometric axes are 60° from the vertical or Y axis. Pick **OK** to accept the settings.

The three isometric projection planes are shown in Figure 8.3 but won't appear on your screen display. The X,Y and Z axes are at 150, 30 and 90 degrees from the horizontal. The orientation of the cursor cross hairs depends on which plane you want to work in. The effect of ORTHO also depends on the plane. The isoplane cube shown in Figure 8.3 defines the planes as LEFT, RIGHT and TOP. You can switch between the planes by pressing ^E or by using the Settings/Drawing aids dialogue box. There is also a command, ISOPLANE, which does this. The orientation of the cursor cross hairs depends on the isoplane setting.

Make sure that ortho is ON (^O) and that coordinate display is set to polar mode. Toggling ^D twice should do this. Alternatively, you can set the value of the system variable, COORDS, to 2. Then switch to the right hand plane to draw the front of the cooker shown in Figure 8.4. You will find it easier to drag the line points than to key them in. Watch the coordinate display for the correct lengths. The exact location of the first point is not too important. However, the relative positions of all other points are. If you make a mistake picking points use the **u** facility in the LINE command to undo that segment.

Command: **COORDS** or use ^D twice
New value for COORDS <1> **2**

Command: **ISOPLANE** or use ^E
Left/Top/Right/<Toggle>: **R**
Command: **LINE**
From point: pick a snap point near (1689,475)   (A)
To point: **@950<90** or drag along the "Iso-Z" axis   (B)
To point: **@500<30** or drag along "Iso-Y" axis   (C)
To point: **@950<270**   (D)
To point: **CLOSE**

The next line, EF, is 100 units above the line DA. The point E is @0,100 from D. The line GH is a further 450 units above EF.

Command: **LINE**
From point: **@0,100**   (E)
To point: **@−500<30**   (F)
To point: **<ENTER>**
Command: **<ENTER>**
LINE From point: **@0,450**   (G)
To point: **@500<30**   (H)
To point: **<ENTER>**

Now switch to the left-hand plane to draw the side. You can use the ISO-PLANE again or use the toggle key ^E. Pressing ^E once changes to the left plane, once again and the top plane is set. You can cycle through all the planes quickly using ^E. This toggle is also transparent so you can switch in the middle of another command.

Command: ^**E** <Isoplane left> **LINE**
From point: **int** of pick point A using object snap.   (A)
To point: **@600<150**   (J)
To point: **@950<90**   (K)
To point: **int** of pick point B.   (B)
To point: **<ENTER>**

Use ^E to toggle to the top plane to finish the cooker.

Command: ^**E** <Isoplane top> **LINE**
From point: **int** of pick point K.   (K)
To point: **@500<30**   (L)
To point: **int** of pick point C   (C)
To point: **<ENTER>**
Command: **<ENTER>**
LINE From point: **@500<150** A point on CL 100 from L   (M)
To point: **@500<210**   (N)

To point: <**ENTER**>

To draw the heating elements you have to distort the circles. Luckily, the
ELLIPSE command is just right for the job. The elements are at 200mm
centers. The center of each circle is 150mm from the nearest edge. First, change
the entity color to red. You will then use the Isocircle option in the ELLIPSE.

Command: **COLOR**
New entity color <BYLAYER>: **RED**
Command: **ELLIPSE** or pick **Draw/Ellipse/Axis, Eccentricity**
<Axis endpoint 1>/Center/Isocircle: **I**
Center of circle: pick the snap point near (1516,1675), 3 snaps from N.
<Circle radius>/Diameter: **75**   (The back left heating ring)
Command: **COLOR**
New entity color <1 red>: **BYLAYER**   (Reset to normal)

AutoCAD uses the isoplane setting to calculate the correct amount of distortion and the orientation of the ellipse. This is a special feature of the ELLIPSE
command, triggered when the SNAP style is isometric. Unfortunately, the AR-
RAY command does not support the isometric planes and so you have to use
the straightforward COPY command.

Command: **COPY**
Select objects: **LAST**
1 found
Select objects: <**ENTER**>
<Base point or Displacement>/Multiple: **M**
Base point: **0,0**   (Any point will do.)
Second point of displacement: **@200<30**   (The back right ring)
Second point of displacement: **@200<−30**   (The front left ring.)
Second point of displacement: <**ENTER**>
Command: **COPY**
Select objects: **LAST**   (The front left ring.)
1 found
Select objects: <**ENTER**>
<Base point or Displacement>/Multiple: **0,0**
Second point of displacement: **@200<30**   (The front right ring.)
Command: **QSAVE**

The absolute X,Y coordinates that appear on the status line don't mean
much when you are working in isometric projection. What is important is the
relative position from the last point. Remember that the lines with <150 are
parallel to the Iso-X axis, those at <30 are supposed to represent the Iso-Y

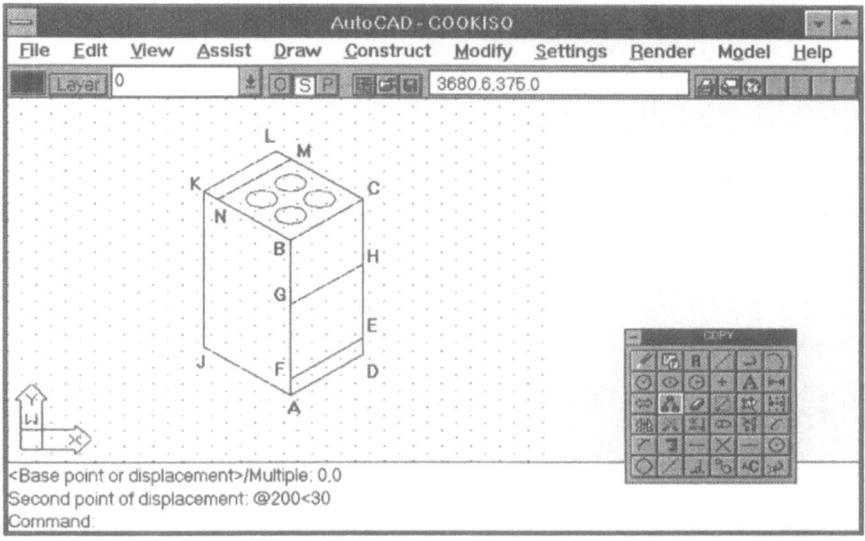

Figure 8.4 An isometric cooker

axis and the vertical direction is Iso-Z. The cooker in Figure 8.4 is only a projection of the 3D information, it is not a 3D object.

The usefulness of AutoCAD's isometric projection will depend on the user's skill in that drafting technique. In general, you will have to use many construction lines to locate key points in the isometric view. Ortho, grid, snap and isoplane are very effective, while typing coordinates is not.

The Express State Building in 2.5 dimensions

It's time for a change of scene for all you out there, slaving over hot stoves. The next stop for the AutoCAD Express is the Big Apple where the skyline is about to be committed to the PC. In this example you will use the conventional 2D operations to draw a plan view. By also assigning a thickness and elevation in the vertical direction (Z axis) the shapes will have body as shown in Figure 8.1.

Pick **File/New** to start a new drawing, calling it "EXP-NY", Use the drawing EXPROTO, that you created above, as the prototype. If you didn't do the previous exercise, start a new drawing and follow the AutoCAD commands given above, down as far as the SAVE "EXPROTO" line.

To give depth to the drawing entities use the Entity Creation Modes dialogue box (Settings/Entity Modes). This allows you, not only to set the

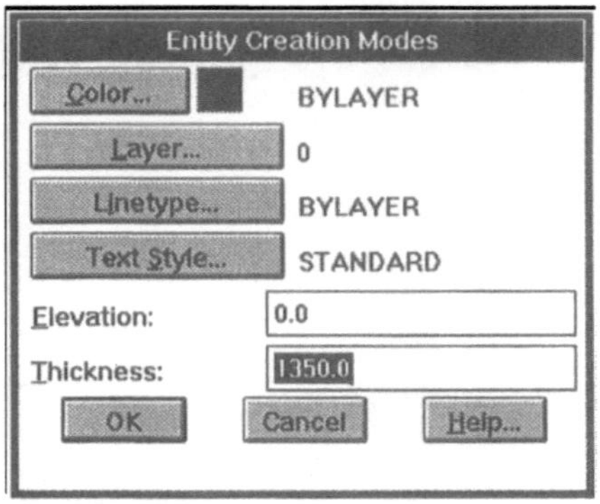

Figure 8.5 Setting thickness

layer, color and linetype but also to set the altitude of the drawing plane and also the thickness or height of the entities(Figure 8.5. The main tower of the building is a massive 500 units by 400 units and has a height, or thickness, of 1350 units. The height to the top of the mast is 2500 units.

Pick **Settings/Entity Modes...** and give a thickness of **1350** and leave the elevation at **0** as shown in Figure 8.5. Then pick **OK**. Setting the elevation to zero means that the base of the tower is at ground level. Now ZOOM in to the construction area and draw a rectangle for the main tower, ABCD in Figure 8.6. Set a SNAP value of 50 and GRID of twice this.

Command: **ZOOM**
All/.../Window/<Scale (X/XP)>: **W**
First corner: **700,700**
Other corner: **1800,1500**
Command: **SNAP**
Snap spacing or ON/OFF/Aspect/Rotate/Style < >: **50**
Command: **GRID**
Grid spacing(X) or ON/OFF/Snap/Aspect < >: **2X**

Now draw the base tower, ABCD,

Command: **LINE**
From point: **1000,1000**   (A)
To point: **@500,0**   (B)
To point: **@0,400**   (C)
To point: **@−500,0**   (D)
To point: **CLOSE**   (A)

The next part up, EFGH in Figure 8.6, is not quite as large but is 500 units tall. Therefore the elevation and thickness must be changed. By setting the elevation to 1350 the new entities will be drawn at the top of the main tower. This can be done using **Settings/Entity Modes...** as above or by using the **ELEV** command.

Command: **ELEV**
New current elevation <0>: **1350**
New current thickness <1350>: **500**
Command: **LINE**
From point: **1100,1000**   (E)
To point: **@300,0**   (F)
To point: **@0,400**   (G)
To point: **@−300,0**   (H)
To point: **C**

As the construction reaches skyward the elevation must be updated for the new elements. Climb to the top of the last object and draw a cylinder. A vertical cylinder is just a circle with a thickness.

Command: **ELEV**
New current elevation <1350>: **1850**
New current thickness <500>: **250**
Command: **CIRCLE**
3P/2P/TTR/<Center point>: **1250,1200**   (J)
Diameter/<Radius>: **150**

And again.

Command: **ELEV**
New current elevation <1850>: **2100**
New current thickness <250>: **100**
Command: **CIRCLE**
3P/2P/TTR/<Center point>: **1250,1200**   (J) again.
Diameter/<Radius>: **100**

Now to draw the mast on top use a POINT with a thickness of 300 units. Of course you must first go to the new elevation. Make sure that the Point Style is set to give dots. Pick **Settings/Point Style...** from the menu bar. The dot should be the top left option in the pop-up.

Command: **ELEV**
New current elevation <2100>: **2200**
New current thickness <100>: **300**
Command: **POINT**

Point: **1250,1200**   (J)

When you are finished working above ground level it is good practice to reset the elevation and thickness back to zero. This will help prevent user confusion if the drawing is made over a number of sessions. You can then add the title text.

> Command: **ELEV**
> New current elevation <2200>: **0**
> New current thickness <300>: **0**
> Command: **DTEXT**
> Justify/Style/<Start point>: **J**
> Align/Fit/Center/...: **C**
> Center point: **1250,850**
> Height <>: **70**
> Rotation angle <0>: **<ENTER>**
> Text: **EXPRESS STATE**
> Text: **BUILDING**
> Text: **<ENTER>**
> Command: **QSAVE**

Note, the centering of the text did not take place until the command input had finished. Your picture should now look like Figure 8.6. If the POINT appears as an X or + or small circle, then you will have to change the Point Style and execute the REGEN command.

Finally, to see the 3D effect of Figure 8.1 you will have to change the view point from which AutoCAD is looking. To get the solid effect you can remove the lines at the back. The commands VPOINT and HIDE do these jobs.

HAZARD WARNING! Always SAVE the drawing before a HIDE operation. The HIDE command can take a long time to calculate all the hidden lines to remove. Don't get impatient and start hitting the <ENTER> key. This only re-executes the last command, ie HIDE, and you will have even longer to wait. Use the cancel key, ^C, if you want to interrupt. As there are only about 25 lines to be hidden in this drawing it shouldn't take more than a few seconds.

Views and more views

The isometric projection view shown in Figures 8.1 and 8.9 is achieved by changing the view point so that we are looking at the tower from an angle. The actual viewing direction is parallel to the line joining this view point to the drawing's TARGET point. The default TARGET point is the origin. Note that the plan view point is (0,0,1), ie looking down the Z axis from that point

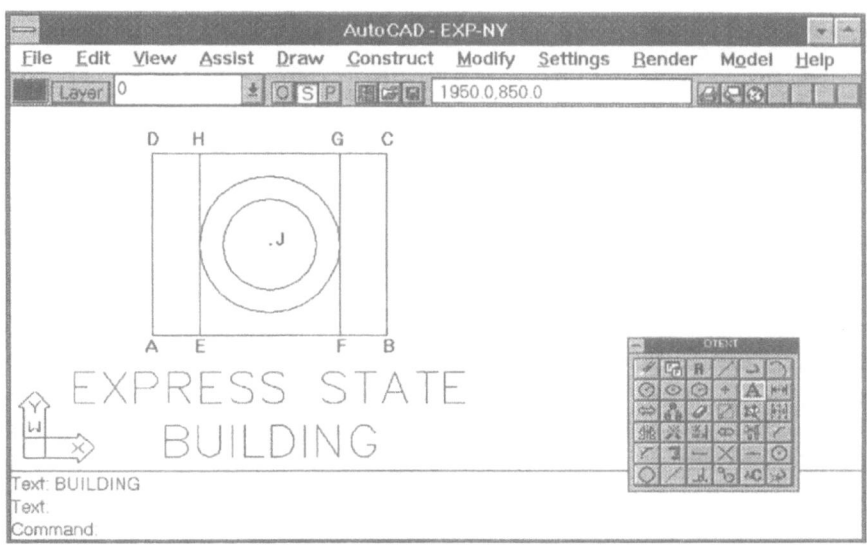

Figure 8.6 Plan view of skyscraper

to the origin. A front elevation of the building could be generated with a view point of $(0,-1,0)$, a back elevation by $(0,1,0)$ and a side view by $(1,0,0)$.

To generate the isometric view you can use the Viewpoint Presets dialogue box. This is obtained by picking **View/Set View** followed by **Viewpoint/Presets**, Figure 8.7. The square with the circle on the left of the dialogue screen represents the X-Y plane. The X-axis is along 0° and the Y along 90°. The graduated semi-circle on the right represents angles from the horizontal plane. A plan view would result from picking 90° on this. The isometric view is generated by moving the cursor into and picking the wedge marked "315°" on the left and the +45° wedge on the right. Alternatively, you could input the angles in the boxes manually. Then pick **OK**.

Command: _DDVPOINT

Now pick the angles **315** and **45**. This calculates the XYZ values for the viewpoint from the angles. Use the quick save command before executing the HIDE command.

Command: **QSAVE**
Command: **HIDE**
Regenerating drawing.
Hiding lines: done 100%

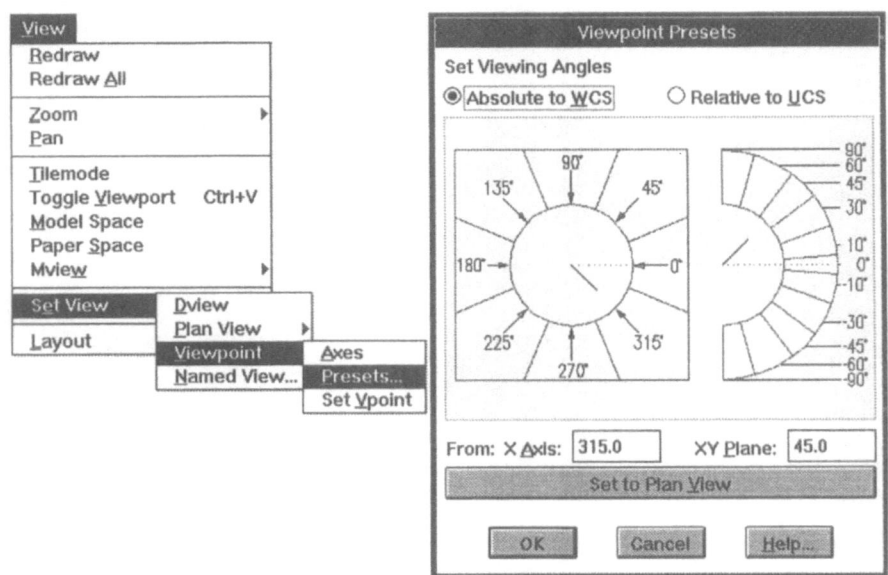

Figure 8.7 Viewpoint presets

This should give the required picture. It is a bit of a fraud, really. If you zoom in to the cylinders on the roof and redo the **HIDE** you should see that one of the lines at the top of the upper rectangular block is not correctly hidden. The reason for this is that the 2.5D lines produce an open ended rectangular box and not a solid block. The thick circles give solid cylinders though.

If you need to store particular view points you can use the **VIEW** command. This allows you to save the current display settings (VPOINT, ZOOM etc) for later retrieval. This is a help when trying to remember the VPOINT coordinates.

Command: **VIEW**
?/Delete/Restore/Save/Window: **S**
View name to save: **ISOMETRIC-P**

The VPOINT command itself can be accessed by picking **View/Set View** from the menu bar followed by **Viewpoint/Set Vpoint**. To store the front view settings, pick **Set Vpoint** and give a value of **0,−1,0**. When you are happy with the display save the VIEW.

Command: **VPOINT**
Rotate/<View point> <0.9,−0.9,1.2>: **0,−1,0**

Command: **VIEW**
?/Delete/Restore/Save/Window: **S**
View name to save: **FRONT**

This could also have been done using the presets of 270° from the X axis and 0 from the XY plane. Using the presets runs the VPOINT command with the "Rotate" option.

When you want to retrieve the view, type **VIEW** followed by **R** and the view name. Try to save a plan view and side view.

Command: **VIEW**
?/Delete/Restore/Save/Window: **R**
View name to restore: **ISOMETRIC-P**

The other views that will be used in Figure 8.9 are the side elevation and the plan. The side view uses a viewpoint of -1,0,0 while the PLAN command gives the latter.

Command: **VPOINT**
Rotate/<View point> <0.9,−0.9,1.2>: **−1,0,0**
Command: **VIEW**
?/Delete/Restore/Save/Window: **S**
View name to save: **SIDE**
Command: **PLAN**
<Current UCS>/Ucs/World: **<ENTER>**

There will be more about VIEW in the next chapter.

Multiple views

It has been possible to show multiple, simultaneous, views since Release 10. In Release 10 these are controlled by the VPORTS command. Since Release 11 uses a more versatile command called MVIEW is available. As the VPORTS command may not be supported in future releases and as it offers nothing that MVIEW cannot do there is not much point in considering it. Furthermore the Release 12 menu system completely ignores the old command. If you have old drawings that use VPORTS you will have to consult *AutoCAD Express 2nd Edition* by yours truly.

Model space and paper space

Drawings like the Express State Building can be considered as models of real objects. There is a direct spatial correspondence between the computer model and the real thing, that is, the model is drawn at full scale. For plotting the model we have to produce a scaled image onto a piece of paper. Indeed, engineering and architectural drawings usually show a number of views of the object on one piece of paper.

Up to now we have created drawings solely in AutoCAD's "model space". We will now set up a virtual "paper space" to view the four images of the skyscraper. Paper space is really a facility for plotting multiple views so the full description of its capabilities is discussed in Chapter 9. However, it also allows us to split the screen into a number of windows which can be very useful in building 3D drawings. In this section only those aspects needed to generate the windows will be described.

The first task in making these windows or metaviews is to set the **TILE-MODE** variable to **0**. This is a transitional variable that allows you to switch between the old VPORTS windows and the newer metaview or MVIEW command. When TILEMODE is 1 or ON a tick will appear in front of it on the View pull-down menu. Pick **View/Tilemode** to switch it off or type the following.

Command: **TILEMODE**
New value for TILEMODE <1>: **0**

The first effect of this is that the skyscraper disappears. Fear not, it hasn't been deleted. Another thing you should notice is that the UCS icon in the lower left corner of the screen changes from X and Y arrows to a triangle. You are now in PAPER SPACE. A P will appear in the status line of the DOS version of AutoCAD while the P button on the toolbar will appear pressed in Windows.

Paper space has its own LIMITS settings which should match the size of paper used for plots. After setting the limits you will define a number of viewports on this virtual page. These can be used to look at the skyscraper from any viewpoint.

Command: **LIMITS**
Reset Paper space limits
ON/OFF/<Lower left corner> <0,0,0>: **<ENTER>**
Upper right corner < >: **420,297**
Command: **ZOOM**
All/.../Window/<Scale(X/XP)>: **A**

The viewports are made using the MVIEW command. The MVIEW sub-menu shown in Figure 8.8 has a number of options. To make the two tall, thin, viewports and the two squarer ones we will use "2 Viewports" twice. Pick **View/Mview/2 Viewports** and follow the prompt sequence below.

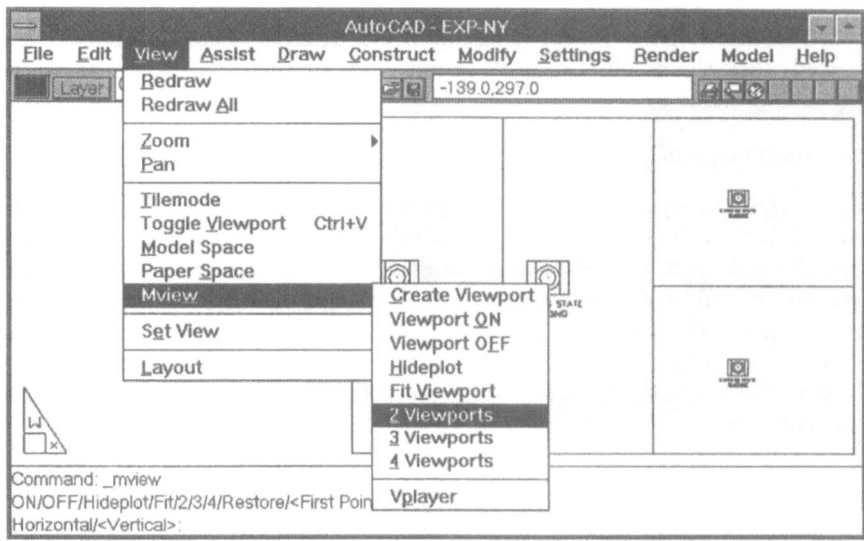

Figure 8.8 Making metaviews

Command: _mview
ON/OFF/Hideplot/Fit/2/3/4/Restore/<First point>: 2
Horizontal/<Vertical>: **<ENTER>**

This means that the two viewports will be split by a vertical line. The following coordinates define the extremities of the two viewports combined. The area will be equally divided between them.

Fit/<First point>: **10,10**
Second point: **210,285**

When a viewport is first defined it will show the active model space view. In this case we left model space with a plan view showing. Now make the two squarer views.

Command: _mview
ON/OFF/Hideplot/Fit/2/3/4/Restore/<First point>: 2
Horizontal/<Vertical>: H
Fit/<First point>: **210,10**
Second point: **410,285**

All that remains now is to go back to model space and select the appropriate view for each viewport. Note that PAPER SPACE is used to define the size and location of the viewport but MODEL SPACE is used to define what is viewed. Pick the **P** button on the toolbar or pick **Model Space** from the **View** pull-down menu.

    Command: MSPACE

The top right viewport should now appear with a heavier outline than the others. This indicates that it is the "active" viewport. If you move the cursor across the screen it will appear as cross hairs only in the active viewport. Elsewhere, it will be a cursor arrow. If the top right viewport is not the active one make it active by moving the cursor into it and press the mouse button.

    To see the plan at a better magnification use the zoom command twice. The first zoom fills the window while the second reduces the size to 90% of the window size.

    Command: **ZOOM**
    All/.../Extents/Window/<Scale(X/XP)>: **E**
    Command: **ZOOM**
    All/.../Window/<Scale(X/XP)>: **0.9X**

Now move to the far left viewport, make it active and restore the SIDE view. You can toggle between viewports by pressing ^**V**.

    Command: **VIEW**
    ?/Delete/Restore/Save/Window: **R**
    View name to restore: **SIDE**
    Command: **ZOOM**
    All/.../Extents/Window/<Scale(X/XP)>: **E**
    Command: **PAN**
    Displacement: **0,−200**
    Second point: <**ENTER**>

The PAN command was used to shift the tower towards the center of the viewport. This may not be necessary. Note that as you move the cursor from left to right in this viewport only the Y value of the coordinates in the status line changes. This is because the X axis is perpendicular to that viewport.

    Now move to the next viewport and restore the FRONT view. Use ^**V** to toggle. When the view is restored only the X values will change in the status line.

    Command: **VIEW**
    ?/Delete/Restore/Save/Window: **R**
    View name to restore: **FRONT**
    Command: **ZOOM**
    All/.../Extents/Window/<Scale(X/XP)>: **E**

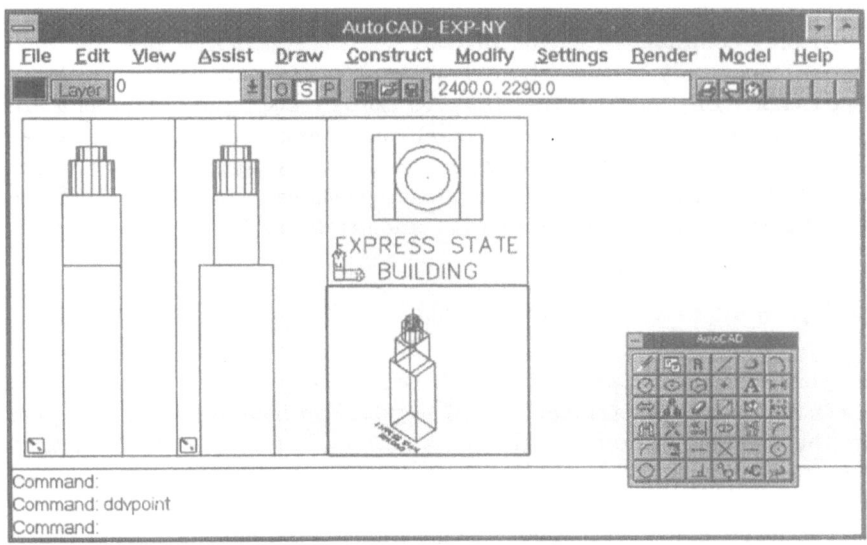

Figure 8.9 Four views of skyscraper

Finally, activate the lower right viewport and restore the ISOMETRIC-P view. Use ZOOM and PAN as required. It is easier to drag isometrics than input exact coordinates. The "1000<45" below moves the tower to the right since that view was defined with an angle of 315° (−45°).

> Command: **VIEW**
> ?/Delete/Restore/Save/Window: **R**
> View name to restore: **ISOMETRIC-P**
> Command: **ZOOM**
> All/.../Extents/Window/<Scale(X/XP)>: **E**
> Command: **PAN**
> Displacement: **1000,<45**
> Second point: **<ENTER>**
> Command: **QSAVE**

Viewports are very useful for 3D CAD by giving you an instant update of new entities in all the views. You can also switch between the viewports for the selection of entities or construction points in the middle of other commands. Their usefulness is not confined to 3D work and they can help speed up 2D drafting considerably. One viewport can be used to show a small-scale picture of the whole drawing, while other viewports can contain various details for working on.

Display commands such as ZOOM, VIEW, REDRAW and coordinate selection apply only to the active viewport. Two commands, REDRAWALL and REGENALL, cause all the viewports to be redisplayed. There is a limit to the number of viewports depending on your system. On PC's this is usually 16. If you want to get rid of a viewport, you must go to paper space and use the ERASE command. There are many more features relating to MVIEW and PAPER SPACE. These are covered in some detail in Chapter 9.

The Pyramids of Giza in glorious 3D

Pack your bags and board the AutoCAD Express for your next destination, the ancient and three-dimensional land of Egypt. You have probably recognised from the previous section that there is a new level of complexity when trying to control points in 3D. In this section you will learn how to master this and construct a fully three-dimensional object, the Cheops pyramid.

The most difficult aspect of working in 3D is the optical illusion you encounter because the screen is only two-dimensional. To help with this problem you can set up a viewport for visualisation and give it a suitable VPOINT. You will also use the coordinate filters, .x, .y, .z, .xy which allow you to pick the x value from one window and the y and z from others.

Pick **File/New** to create a new drawing called **EXP-GIZA** using **EX-PROTO** as the default. Set SNAP to 50, GRID to 100 and draw the plan for the pyramid shown in Figure 8.10. After that, use MVIEW to set up three viewports on the screen to watch the 3D take off.

```
Command: SNAP
Snap spacing or ON/OFF/Aspect/Rotate/Style < >: 50
Command: GRID
Command: LINE
From point: 1000,1000                                              (A)
To point: @600,0                                                  (B)
To point: 0,600                                                   (C)
To point: −600,0                                                  (D)
To point: C                                                       (A)
Command: TILEMODE
New value for TILEMODE <1>: 0
```

Now set the limits for paper space to (420,297) as before and make the two left-hand viewports shown in Figure 8.10.

```
Command: LIMITS
Reset Paper space limits
ON/OFF/<Lower left corner> <0,0,0>: <ENTER>
```

Upper right corner < >: **420,297**
Command: **ZOOM**
All/.../Window/<Scale(X/XP)>: **A**
Command: **MVIEW**
ON/OFF/Hideplot/Fit/2/3/4/Restore/<First point>: **2**
Horizontal/<Vertical>: **H**
Fit/<First point>: **210,10**
Second point: **410,285**

Now make the larger viewport appear on the right-hand half of the screen. This will also be the new active viewport when you go to MODEL SPACE. Once in model space, use ZOOM W to make better use of the display.

Command: **MVIEW**
ON/OFF/Hideplot/Fit/2/3/4/Restore/<First point>: **210,10**
Other corner: **410,285**
Command: **MSPACE**
Command: **ZOOM**
All...//Window/<Scale (X/XP)>: **W**
First corner: **700,700**
Other corner: **2000,1800**

Make the lower left viewport active by moving the cursor into it and pressing the pick button. Then change the VPOINT to give a frontal view. Move the the upper left and set up an isometric type of view.

Pick lower left viewport.
Command: **VPOINT**
Rotate/<View point> <0,0,1>: **0,−1,0**
Regenerating drawing.
Command: **ZOOM**
All...//Window/<Scale (X/XP)>: **0.7X**
Pick upper left viewport.
Command: **VPOINT**
Rotate/<View point> <0,0,1>: **1,−1,1**
Command: **ZOOM**
All...//Window/<Scale (X/XP)>: **0.8X**

With these views you should be able to see if the lines to the apex of the pyramid are being drawn correctly. There are four sloping lines to be drawn from the corners, A,B,C and D, to the apex, E which is 475 units above the center point. To explain the facilities a number of coordinate definition methods are used.

> Command: **LINE**
> From point: **1000,1000**   (A)
> To point: **@300,300,475**   (E)

If the Z ordinate is not specified it is taken as the current elevation setting. The point E is 300 units along the X axis, 300 along the Y axis and 475 units up the Z axis from the point A. As the second point is input the line should appear in all three viewports. For the next point use object snap intersection to locate B in the right hand viewport.

> To point: **intersec**
> of Pick the right hand viewport and then pick the point B.
> To point: **<ENTER>**

The object snap is also able to pick up the full XYZ coordinates as demonstrated by the next sequence.

> Command: **<ENTER>**
> LINE From point: **intersec**
> of Pick point C   (C)
> To point: **intersec**
> of Pick point E.   (E)

If you are not sure of the elevation of a particular point on the plan view you can use the .XY filter to use just those coordinates and type in Z separately.

> To point: **.xy**
> of **intersec**
> of Pick the point D.   (D)
> (Need Z): **0**
> To point: **<ENTER>**

The complete pyramid should now appear similar to Figure 8.9 but without the letters A to E. You may need to do a Zoom, Extents in the two lefthand viewports.

Making the faces solid

The pyramid shown above is but the first step in construction. The 3D lines form only the frame on which we can hang the fabric. To make the slopes solid we will create 3DFACEs on each of the triangles. Making the 3D faces is like stretching fabric over the wire frame. Faces are opaque when the HIDE command is executed. At present, if you move to the upper left viewport and try HIDE, all the lines will still be visible.

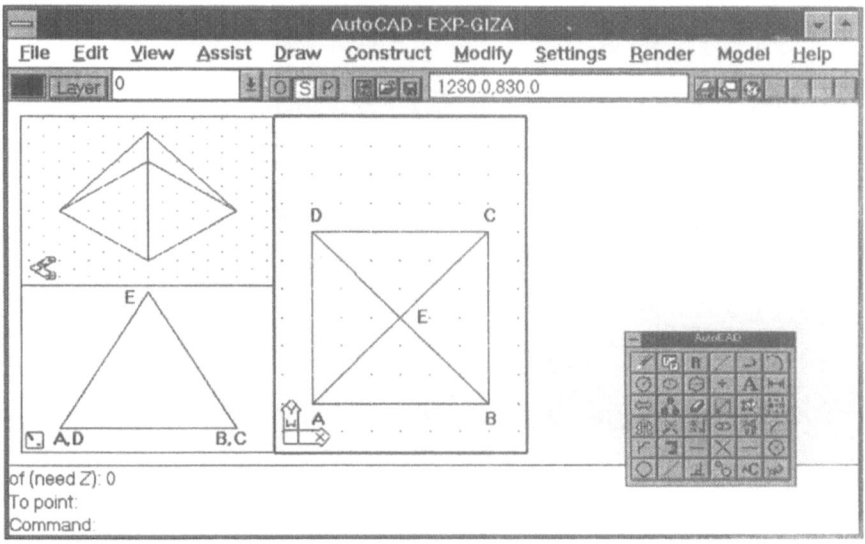

Figure 8.10 Cheops' pyramid

Create a new layer with a different color so that the faces are distinguishable from the original lines. It will be helpful later if the floor of the pyramid is on a separate layer from the walls. Make sure that the right hand viewport is active before starting on the faces. Then make a 3D face for the bottom of the pyramid and a face for each of the other four sides.

Command: **LAYER**
?/Make/Set/...: **N**
New layer name(s): **FACES,FLOOR**
?/Make/.../Color/...: **C**
Color: **red**
Layer name(s) for color 1 (red) <0>: **FACES,FLOOR**
?/Make/Set/...: **S**
New current layer <0>: **FLOOR**
?/Make/Set/...: **<ENTER>**

The 3DFACE command is found by picking **Draw/3D Surfaces** from the menu bar (Figure 8.11). Now pick **3DFACE**. We will return to the other items on the 3D menu later. The first face is to be the floor of the pyramid.

Command: 3DFACE
First point: **1000,1000**   (A again.)

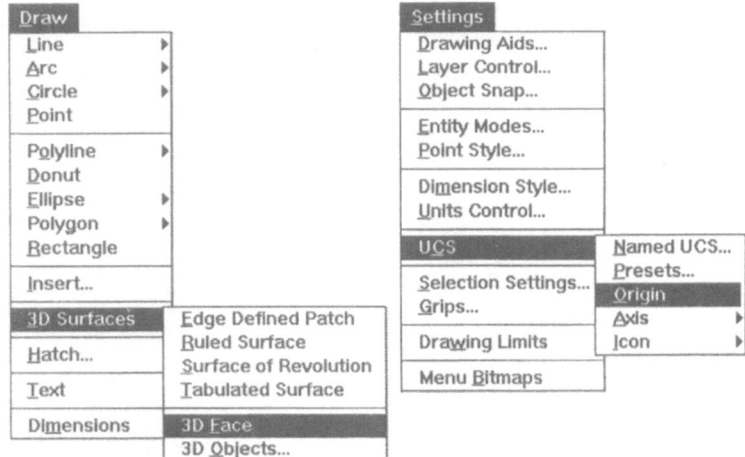

Figure 8.11 3D and UCS menus

Second point: **@600,0,0**   (B)
Third point: **@600<90**   (C)

Let's use the filters for the last point, for fun. Point D has the same X value
as A and the same Y as C.

Fourth point: **.x**
of **int** of Pick point A.
(need YZ): **.y**
of **int** of Pick point C.
(need Z): **0**
Third point: **<ENTER>**

You will be prompted for more third and fourth points to add more faces
onto the last edge. In fact, the 3DFACE command is like a 3D version of the
SOLID command. However, the order of point input (ABCD above) is more
coomprehensible than that for SOLID (ABDC). Note that only the edges of the
face are shown. Faces are never filled but they are opaque when using HIDE.
The four points defining the face should be on the same plane if possible. It is
not an error to use non-coplanar points but it is sloppy 3D CAD. The picture
will look the same as Figure 8.10 but the square should now be in red.

Now make the sloping faces ABE and CDE in one 3DFACE bow-tie operation. This will be followed by BCE and ADE. First, switch layers to FACES.

```
Command: LAYER
?/Make/Set/...: S
New current layer <0>: FACES
?/Make/Set/...: <ENTER>
Command: 3DFACE
First point: 1000,1000                                    (A)
Second point: @600,0                                      (B)
Third point: intersec of Pick point E                     (E)
Fourth point: <ENTER>
Third point: intersec of Pick point C                     (C)
Fourth point: intersec of Pick point D                    (D)
Third point: <ENTER>
Command: <ENTER>
3DFACE First point: 1000,1000                             (A)
Second point: @0,600                                      (D)
Third point: intersec of Pick point E                     (E)
Fourth point: <ENTER>
Third point: intersec of Pick point C                     (C)
Fourth point: intersec of Pick point B                    (B)
Third point: <ENTER>
Command: QSAVE
```

Faces always have four points defining them. To make a triangle two of the points must have the same location (eg first and second or third and fourth). Note that the third and fourth points of the previous face are used as the first two points in the next.

Your pyramid should look much the same as before, but this time in red. To see the difference between the 3DFACE representation and the LINEs, move to the upper left viewport and issue the HIDE command. Then make the layer containing the lines the current layer and freeze layer, FACES. Try HIDE once more and you should still be able to see all the lines. Thaw and Set the FACES layer again for the next part of the exercise.

Define your own coordinate system

The most important advance in 3D AutoCAD has been the introduction of user definable coordinate systems. This means that you can reset the position of the origin and also the orientation of the X, Y and Z axes. The default coordinates system that has been used up to now is the World Coordinate System (WCS). The WCS specifies the drawing origin and the directions of X, Y and Z axes. Other coordinate systems are defined relative to this.

One possible point of difficulty can be deciding on which direction the positive Z axis points towards. AutoCAD uses the right hand rule to define all coordinate systems. Place your right hand near the computer screen with your palm facing you and extend the thumb to the right, forefinger up and middle finger towards you. These fingers show the positive directions of the X, Y and Z axes respectively. If you keep your fingers in that postiion and rotate your hand you will see how the axes of the new coordinate system relate to each other.

In this section you will draw an inscribed circle on the ABE slope. To try this in the WCS would be fruitless because AutoCAD circles are always drawn in the XY plane. You have to define a User Coordinate System (UCS) parallel to the slope. In fact you have to make a new UCS for every new plane you want to draw circles or other 2D entities in.

The UCS command appears on the Settings pull-down menu. Pick **Settings/UCS** to see the menu shown in Figure 8.11. Pick **Origin** to reset the origin to the point A on the pyramid.

> Command: **UCS**
> Origin/Zaxis/.../?/<World>: **O**
> Origin point <0,0,0>: **1000,1000,0**   (A)

If you pick **Settings/UCS** followed by **Icon/Origin** the little UCS icon should move to A in the active viewport. If you look closely, you will see that the "W" has disappeared from the icon. A small "+" indicates that it is at the UCS origin. This shifting of the origin can be very useful even in 2D drawings. Note that all coordinate values are now relative to this new origin.

Now define the four slopes as new coordinate systems. The options for defining the plane are quite varied. You can align the UCS with an entity such as a 3DFACE, or you can specify three points in the plane. You can also select the XY plane or specify a new Z axis direction. The UCS can also be set to a particular VIEW or rotated about any of the XYZ axes. Different UCS definitions can be named and saved and restored like VIEWs.

> Command: **UCS**
> Origin/.../Entity/.../Save/Del/?/<World>: **E**
> Select object to align UCS: Pick the face ABE along edge AB.

The UCS icon should now take up its new orientation. Save this UCS as "ABE". The position of the origin is dependent on which point of the 3D face was originally drawn first. This method cannot be used with entities that contain non-coplanar points.

Now save this UCS. Pick **Settings/UCS** and **Named UCS....** The UCS control dialogue box then pops up (Figure 8.12). The UCS that has just been defined is the current one. It appears on the list as "*NO NAME*". To give it the name "ABE" pick the line, "*NO NAME   Cur" and type ABS in the

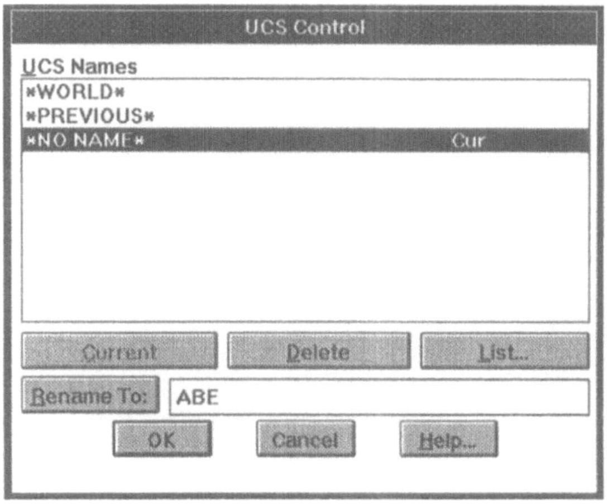

Figure 8.12 UCS Control

box near the bottom as shown in Figure 8.12. Then pick the **Rename To:** button followed by **OK**.

Define the BCE plane by picking each of the three points with object snap intersection and save it.

Command: **UCS**
Origin/Zaxis/3point/...<World>:**3**
Origin point <0,0,0>: **INT** of Pick point B.
Point on positive portion of X axis <601,0,0>: **INT** of Pick C.
Point on positive-Y portion of the UCS XY plane <599,0,0>:
 INT of Pick E.
Command: **UCS**
Origin/.../Save/Del/?/<World>:**S**
?/Desired UCS name: **BCE**

To demonstrate this method further restore the WCS and define a new UCS for the side CDE. Remember the right hand rule for positive axis directions. Pick **Settings/UCS/Presets**. Then pick the World coordinate system icon, **top left**, from the UCS Orientation pop-up (Figure 8.13). This pop-up also gives quick access to the previously used UCS and allows you to set the current view as a UCS. The other presets are useful for swopping from the front to the back of an object or from the front to the side. These presets, however, are better for rectangular buildings than for the pyramid.

You can also set the world coordinate system with the UCS command.

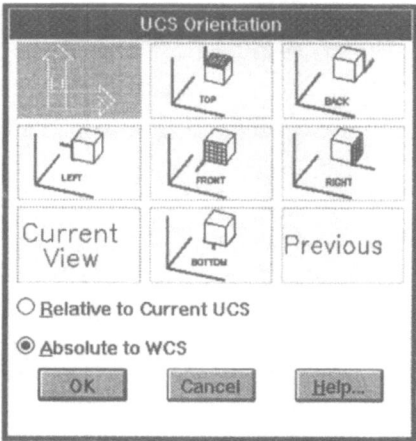

Figure 8.13 UCS Orientation

Command: **UCS**
Origin/.../Save/Del/?/<World>:**W**
Command: **UCS**
Origin/ZAxis/3point/...<World>:**3**
Origin point <0,0,0>: **1600,1600,0**   (C)
Point on positive portion of X axis <1601,1600,0>:
    1000,1600,0   (D)
Point on positive-Y portion of the UCS X-Y plane <1600,1599,0>:
    1300,1300,475   (E)
Command: **UCS**
Origin/.../Save/Del/?/<World>:**S**
?/Desired UCS name: **CDE**

Finally, use the Entity option to make the fourth UCS for side DAE.

Command: **UCS**
Origin/.../Entity/...<World>:**E**
Select object to align UCS: Pick the face DAE along edge AD.

The UCS icon moves to point A since that was the first point used to originally
draw the 3D face. The X axis is positive along the line AD as that was the
original input order of the points. The positive Y axis is up the face. From the
right hand rule this means that the Z axis is positive into the pyramid. All the
other faces have UCS's with Z positive out from the pyramid. To make this last
UCS consistent with the others, you should move the origin to D and make D

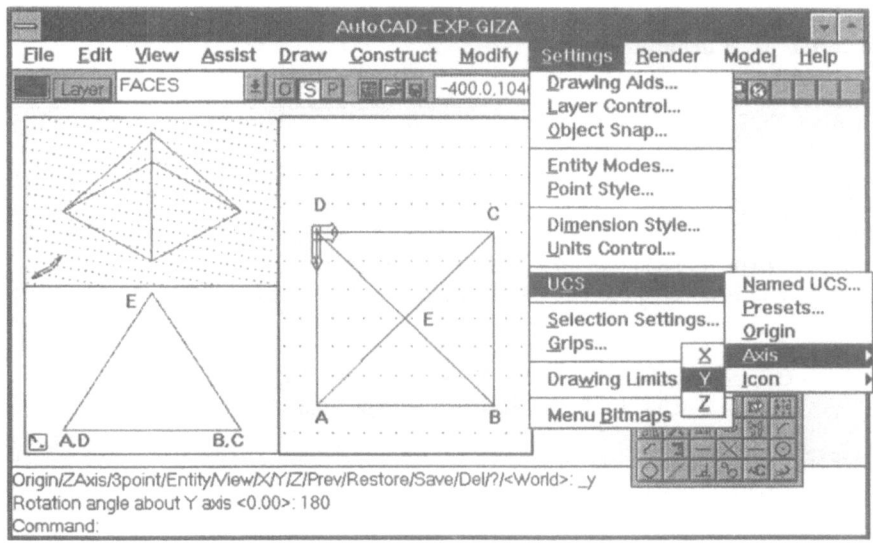

Figure 8.14 Rotating about Y-axis

to A the positive direction for the X axis. This latter task can be accomplished by rotating the UCS about the Y axis. Pick **Settings/UCS/Origin** to shift the origin 600 units along the X- axis.

```
Command: _ucs
Origin/.../X/Y/Z/...<World>:O
Origin point <0,0,0>: 600,0                    Point D in current UCS.
```

Now pick **Settings/UCS/Axis/Y** to rotate the X-axis 180° about the Y-axis to give the situation shown in Figure 8.14. This effectively flips the UCS over. You should now see the icon at D with the X arrow pointing towards A and Y pointing up the slope. Then save DAE.

```
Command: _ucs
Origin/.../X/Y/Z/...<World>: _y
Rotation angle about Y axis <0.0>: 180
Command: UCS
Origin/.../Save/Del/?/<World>:S
?/Desired UCS name: DAE
```

To get a detailed list of these defined coordinate systems use the "?" option from the UCS command:

Table 8.1 UCS definitions

Current UCS: DAE

Saved coordinate systems:

ABE

       Origin = <600,−0,−0>, X axis = <0,0.5,−0.8>
       Y axis = <−0.5,0.7,0.4>, Z axis = <0.8,0.4,0.3>

BCE

       Origin = <600,320,−507>, X axis = <−1,−0,0>
       Y axis = <0,0.4,0.9>, Z axis = <−0,0.9,−0.4>

CDE

       Origin = <−0,320,−507>, X axis = <0,−0.5,0.8>
       Y axis = <0.5,0.7,0.4>, Z axis = <−0.8,0.4,0.3>

DAE

       Origin = <0,−0,0>, X axis = <1,0,−0>
       Y axis = <0,1,0>, Z axis = <0,−0,1>

    Command: **UCS**
    Origin/. . ./Save/Del/?/<World>:?
    UCS names to list <*>: **<ENTER>**
    The coordinates in the "< >" in Table 8.1 are relative to the current UCS.
To see the definitions with respect to the WCS you will first have to select the
W option and repeat the above command.

    Origin/. . ./Prev/Restore/Save/Del/?/<World>: **W**
    You can restore any of the named UCS's, or delete ones that are no longer
required. The "Prev" option restores the previous UCS setting similarly to the
UCS Orientation pop-up.

The all seeing eye

You can now use these four coordinate systems to add bricks to the pyramid
walls and to draw the "all seeing eye". Use the right hand viewport for the
following constructions. If it is not already the active viewport then move the
cursor into the right hand viewport and press the mouse button.

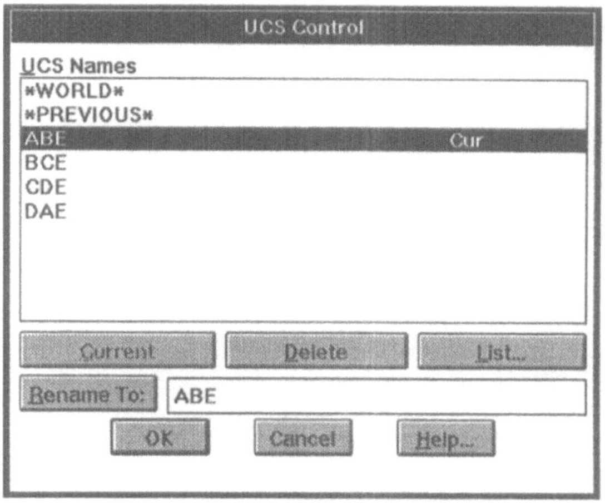

Figure 8.15 Setting UCS to ABE

Pick **Settings/UCS/** and **Named UCS...** to get the UCS Control dialogue box shown in Figure 8.15. Then pick **ABE** from the list followed by the **Current** button and **OK.** To set up a plan view before drawing the eye, pick **View/Set View** followed by **Plan View/Current UCS** as shown in Figure 8.16.

This causes the view point to change so that you are now looking perpendicularly down on the face ABE. AutoCAD executes a ZOOM Extents automatically when a PLAN has been selected. This means that the edge lines will appear at the very edge of the viewport. To get a better picture use **ZOOM 0.9X** to reduce the size. Then create two new layers on which to draw the wall pictures.

Command: **ZOOM**
All/.../Window/<Scale (X/XP)>: **0.9X**
Command: **LAYER** Or use Settings/Layer Control
?/Make/Set/New/...: **N**
New layer name(s): **EYE,WALLS**
?/Make/.../Color/...: **C**
Color: **BLUE**
Layer name(s) for color 5 (blue) <FACES>: **EYE**
?/Make/Set/...: **S**
New current layer <FACES>: **EYE**
?/Make/Set/...: **<ENTER>**

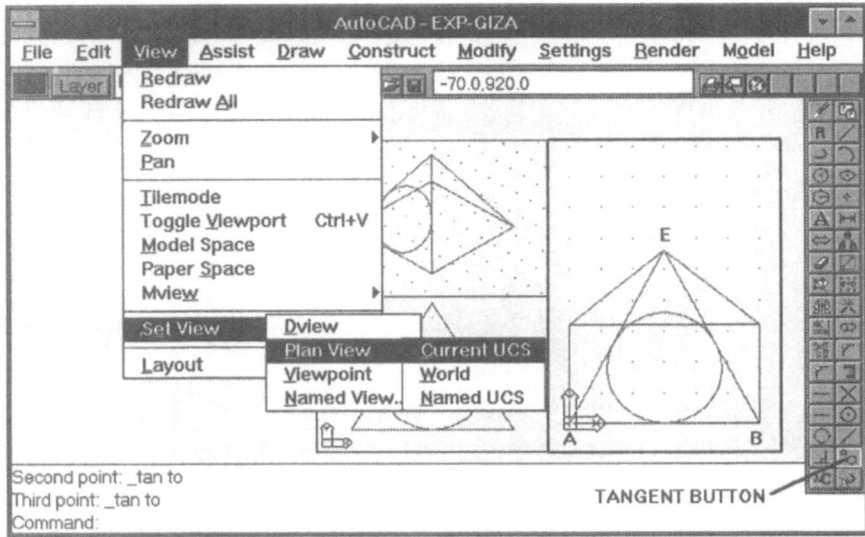

Figure 8.16 3 Point circle

To inscribe a circle in the ABE triangle pick **Draw/Circle/3 Point** and use object snap "TANgent" for each point. Note that the tool box in Figure 8.16 is fixed on the right. This was achieved clicking the tool box icon in the toolbar three times.

> Command: **CIRCLE**
> 3P/2P/TTR/<Center point>: **3P**
> First point: **TAN** to Pick the line AB.
> Second point: **TAN** to Pick the line AE.
> Third point: **TAN** to Pick the line BE.

The circle should fit nicely in the triangle. You can now use two arcs to draw the eye shape. Use the quadrant points of the circle with object snap to locate the end and center points. Then mirror the arc about the line **FG** (Figure 8.17).

> Command: **ARC**
> Center/<Start point>: **QUAD** of Pick circle near point F.
> Center/End/<Second point>: **E**
> End point: **QUAD** of Pick circle on its left, near point G.
> Angle/Direction/Radius/<Center point>: **300,0**
> Command: **MIRROR**

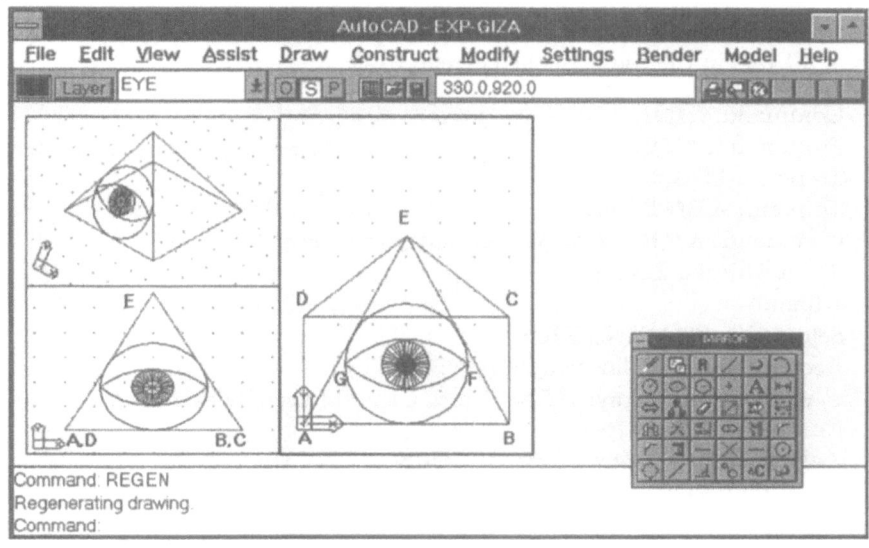

Figure 8.17 The all seeing eye

Select objects: **LAST**
1 found
Select objects: **<ENTER>**
First point of mirror line: **INT** of Pick point F.
Second point: **@1,0**   (A horizontal line.)
Delete old objects? **<N>:** **<ENTER>**

The pupil and iris will complete the all seeing eye. A doughnut with no hole makes a good pupil while a circular array gives the iris. Another circle encloses both. Use the **Draw** pull down menu or type the commands.

Command: **CIRCLE**
3P/2P/TTR/<Center point>: **CEN** of Pick the inscribed circle.
Diameter/<Radius>: **70**
Command: **DONUT**   (Can also be spelt "DOUGHNUT".)
Inside diameter <0.5>: **0**
Outside diameter <1>: **60**
Center of doughnut: **@**   (The center of the circles.)
Center of doughnut: **<ENTER>**

You are prompted to place more donuts. Pressing **<ENTER>** exits the command. Now draw the line from the doughnut to the small circle and ARRAY it. Array is on the **Construct** menu. While this is a 3D drawing, this array is only a 2D one in the plane of the current UCS.

```
Command: LINE
From point: @30,0                    (30 to the left of the donut's center.)
To point: @40,0
To point: <ENTER>
Command: ARRAY or pick Construct/Array
Select objects: LAST
1 found
Select objects: <ENTER>
Rectangular or Polar array (R/P): P
Center point of array: CEN of pick either of the circles.
Number of items: 36
Angle to fill (+ =ccw, − =cw) <360>: <ENTER>
Rotate objects as they are copied? <Y>: <ENTER>
```

The all seeing eye of Cheops' pyramid should now look like that in Figure 8.17. The donut will appear fully solid only in plan views. Now is a good time to SAVE the drawing as the next task is to HATCH the walls with a brick pattern. There is a potential problem in selecting the faces to hatch. All the red lines on the drawing belong to two faces. If you select a face by picking its edge you may get the adjacent face instead. To avoid this you can freeze the FLOOR layer and pick each of the wall faces using the bottom line.

```
Command: QSAVE
Command: LAYER or pick the menu Settings/Layer Control.
?/Make/Set/...: S
New current layer <EYE>: WALLS
?/Make/.../Freeze/Thaw...: Freeze
Layer name(s) to freeze: FLOOR
?/Make/Set/...: <ENTER>
```

The hatching operation firstly involves selecting the pattern, the scale and the hatching style. Pick **Draw/Hatch**... to get the Boundary Hatch pop-up screen (Figure 8.20). Pick **Hatch Options**... and pick the **Pattern**... from the options dialogue box (Figure 8.19). Press **Next** on the pattern pop-up to locate the one called **brick**, second last on the bottom row in Figure 8.18. Then pick the rectangle of bricks just above the word "brick". The screen reverts to the Hatch Options dialogue box. Having selected the pattern, you must now specify the scale as **4.0** and check that the style is normal, as shown in Figure 8.19. When everything is satisfactory pick **OK** to return to Figure 8.20.

Now click the **Select Objects** button and pick the face, ABE, along the bottom line and the larger circle.

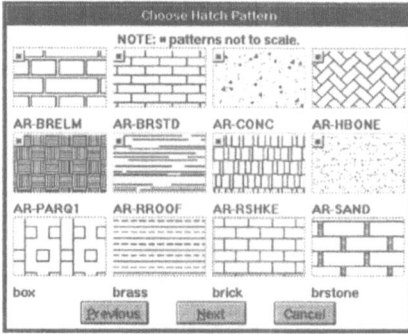

Figure 8.18 Selecting brick pattern

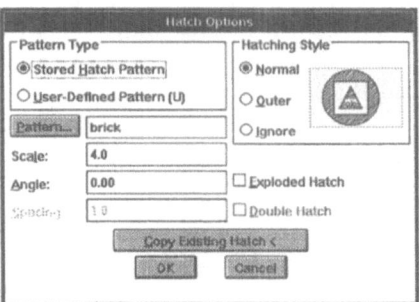

Figure 8.19 Hatch options

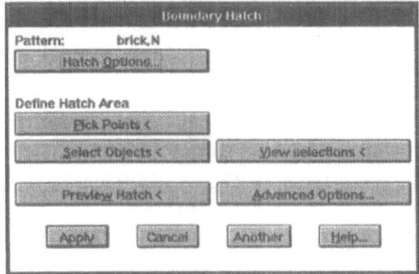

Figure 8.20 Boundary hatch pop-up

Select objects: **300,360** or pick the larger circle.
1 found
Select objects: **140,0** or pick the face ABE along the line AB.
1 found
Select objects: <**ENTER**>

AutoCAD returns to Figure 8.20 for further instructions. Before executing
the hatch, pick **Preview Hatch** to see what it will look like. If it looks like
the lower left view in Figure 8.21 then press <**ENTER**> and pick **Apply**.
Otherwise, go through the above procedure to fix any errors.

Press RETURN to continue. <**ENTER**>
Now pick **Apply**

The area outside the circle should now be hatched in all three viewports.
The hatch only becomes part of the drawing when the Apply button has been
picked. If the wrong area has been bricked in use "ERASE, last" and try again.

To put the bricks on the other three walls you will have to restore each
UCS and execute the HATCH command. Thus, you now need to use the UCS
command to restore the BCE coordinate system.

Command: **UCS**
Origin/.../Restore/.../<World>: **R**
?/Name of UCS to restore: **BCE**

Now execute the brickwork by typing the HATCH command.

Command: **HATCH**
Pattern (? or name/U,style) <BRICK>: <**ENTER**>
Scale for pattern <4.0000>: <**ENTER**>
Angle for pattern <0.00>: <**ENTER**>
Select objects: **140,0** or pick the face BCE along the line BC.
1 found
Select objects: <**ENTER**>

The bricks should now appear on the right hand wall. Note that even though
the object is being viewed from the front, the hatch is always applied in the
plan view and current elevation of the *CURRENT UCS*. This can be a bit
disconcerting and often leads to errors. Therefore is is recommended that you
execute a PLAN command after changing the UCS.

This is easier done than said, using the system variable, USCFOLLOW.
Giving it a value of 1 causes AutoCAD to go automatically to the plan view
whenever a UCS is restored or newly defined. Setting UCSFOLLOW to 1
enables this feature, 0 disables it. Some versions of the menu have a **Follow**

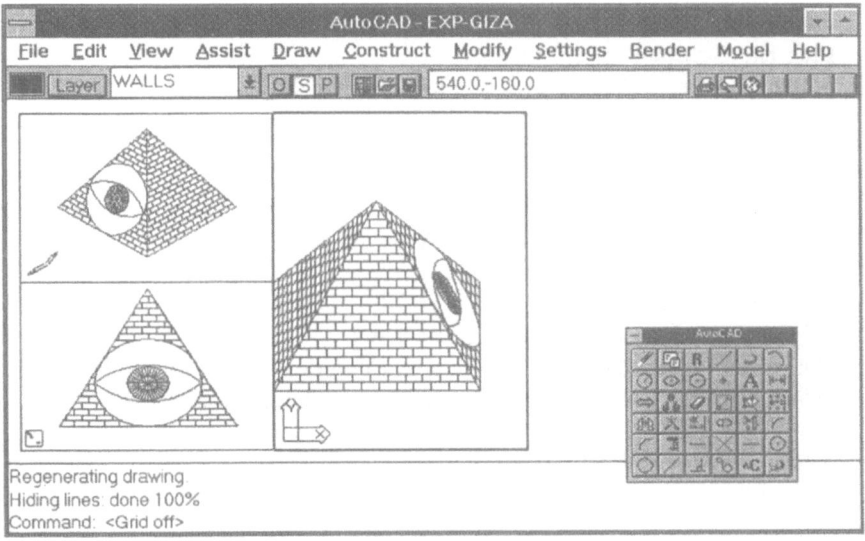

Figure 8.21 The brick-built pyramid

option in the Plan View sub-menu but it is missing in other versions. You can always type it at the command line.

Command: **UCSFOLLOW**
New value for UCSFOLLOW <0>: **1**

Now do the remaining two sides. Note how the display goes straight to the plan view. Type the UCS command or pick **Settings UCS** and **Named UCS....** Then restore CDE by picking it from the list and clicking the **Current** button followed by **OK**. The next sequence is the equivalent set of keystrokes. Do a ZOOM to shrink the image. This will make it easier to pick the bottom line of the face.

Command: **UCS**
Origin/.../Restore/Save/Del/?/<World>:**R**
?/Name of UCS to restore: **CDE**
Command: **ZOOM**
All/.../Window/<Scale (X/XP)>: **0.9X**

This time apply the hatch using the Boundary Hatch pop-up. Pick **Draw/ Hatch....** All the settings from the previous hatch are still valid, so you can select the object, preview and then apply the hatch.

Command: _bhatch
Pick the **Select Objects** button.
Select objects: **140,0** or pick the face CDE along the line DE.
1 found
Select objects: **<ENTER>**
Pick **Apply**
Command: **UCS**
Origin/.../Restore/Save/Del/?/<World>:**R**
?/Name of UCS to restore: **DAE**
Command: **ZOOM**
All/.../Window/<Scale (X/XP)>: **0.9X**
Command: **BHATCH**   (You don't have to use the menu.)
Select objects: **140,0** or pick the face DAE along the line DA.
1 found
Select objects: **<ENTER>**
Pick **Apply**

If you execute a HIDE on the upper left viewport, your picture should look like Figure 8.21. Always save before using HIDE as it can take some time.

Pick the upper left viewport.
Command: **QSAVE**
Command: **HIDE**
Regenerating drawing.
Hiding lines: done 100%

A dynamic view point on visualisation

Another of the goodies included since Release 10 is the Dynamic View command, DVIEW. This is quite an advance on the VPOINT command as it allows you to see the object as you move and twist it in full 3D. DVIEW effectively combines ZOOM, VPOINT, and a perspective view option with a powerful user interface. The emphasis is on 3D visualisation and much of the terminology comes from photography. To use DVIEW you have to imagine yourself looking through a camera lens at a target point.

To see the pyramid in all its glory let's use the full screen and revert to the World Coordinate System and TILEMODE, on. It is advisable to do a zoom such that the object to be viewed appears near to the center of the screen at a low magnification. If the object is near the edge or fills the screen the dynamic view may cause it to go off screen completely.

Command: **TILEMODE**

New value for TILEMODE <0>: **1**
Command: **UCS**
Origin/.../Restore/Save/Del/?/<World>:**W**

Back in tilemode you should see the pyramid in plan. This was the view when tilemode was originally turned off. If your pyramid is not in plan then execute the PLAN command. It is also necessary to thaw the FLOOR layer.

Command: **LAYER**
?/Make/.../Freeze/Thaw...: **THAW**
Layer name(s) to thaw: **FLOOR**
?/Make/Set/...: **<ENTER>**
Command: **PLAN**
<Current UCS>/Ucs/World: **W**
Command: **ZOOM**
All/.../Window/<Scale (X/XP)>: **A**

Now pick **View/Set View** and **Dview** menu bar. You are then prompted to select the objects to be viewed. Here you will select everything. Once the selection has been completed you can begin to explore the many Dview options for manipulating the display.

Command: **DVIEW** Pick from the screen menu.
Select objects: **W**
First corner: **900,900**
Second corner: **1700,1700**
58 found.
Select objects: **<ENTER>**

Only the selected objects will be shown in the dynamic previews. When the final view has been chosen, all the drawing will be displayed. You could, for example remove the hatching from the selection set, which would speed things up slightly.

The command prompt changes to give all the display options. The display will be calculated relative to a given camera position and the target position. The target is the point where the camera is focussed on and will always end up in the center of the screen. The camera can be placed anywhere in 3D space either inside or outside the pyramid. To find the current target position select the **POints** option.

CAmera/TArget/Distance/POints/PAn/Zoom/TWist/CLip/
   Hide/Off/Undo/<eXit>: **PO**
Enter target point < >: **1300,1300,0**
Enter camera point < >: **1300,900,237**

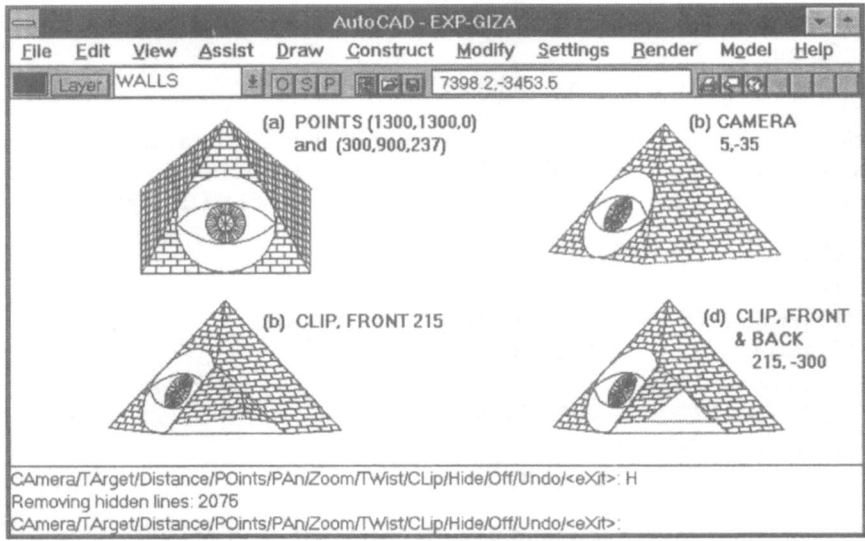

Figure 8.22 Dviews of Cheops with HIDE

This puts the target at the center of the pyramid base and the camera in front of and looking down on the eye (Figure 8.22a). That was a static type of operation. To use the dynamic view select the "CAmera" option. This allows you to specify the angle from the XY plane (the base **ABCD** of the pyramid). A positive angle puts the camera above the target, a negative angle below. A plan view can be generated by using an angle of 90 degrees. You are then asked to put in a camera direction angle relative to the X axis (line **AB**). This angle rotates the camera in a horizontal plane while keeping it focussed on the target point.

When you select the **CA** option you are prompted for new angles with the current values as defaults. You can enter the angle by typing a value or by moving the cross hairs on the screen. As you move the cursor up and down, the angle is shown in the coordinate read-out on the toolbar. The pyramid will be rotated and shown ghosted, in preview mode. The preview image is shown in only one color and is updated continuously. The speed of the update will depend on the number of objects in the dview selection set. Make sure that SNAP is OFF or the action may appear jumpy (Use **^B**). Note that the angle from the XY plane changes with the up–down motion of the cross hairs. Moving the cross hairs from left to right changes the angle from the X axis.

CAmera/TArget/Distance/POints/PAn/Zoom/TWist/CLip/
  Hide/Off/Undo/<eXit>: **CA**
Toggle angle in/Enter angle from XY plane <30.65>: **T**

Pressing T toggles the angle to "from X axis". The angle in the toolbar will now respond to left–right movements of the mouse.

Toggle angle from/Enter angle from XY plane from X axis < −90>: **-35**
Toggle angle in/Enter angle from XY plane <30.65>: **5**

If you use the mouse to pick values both values will be input simultaneously based on the cross hairs' position. Using the toggle and the arrow keys on the keyboard is best. When a suitable angle is found, then key in the value. Once one value has been keyed in then the mouse position will only be used for the remaining angle.

The TArget option works very like CAmera except that it is the target point that moves relative to the camera position. Remember, the target point will always end up in the middle of the screen. If you move the target up then the objects will appear lower in the display. Pressing <ENTER> below accepts the default without any changes.

CAmera/TArget/Distance/POints/PAn/Zoom/TWist/CLip/
  Hide/Off/Undo/<eXit>: **TA**
Toggle angle in/Enter angle in X-Y plane <−5.00>:
  Move the cursor around and then press <**ENTER**>
Toggle angle from/Enter angle in X-Y plane from X axis <145.00>:
  Use slider and press <**ENTER**>

Dview's PAn and Zoom options work similarly to the normal commands. The zoom is, however, a restricted version of the normal command. It only allows you to change the magnification, similar to the ZOOM with Scale(X). You get a slider bar so you can see the effect before picking a scale factor (Figure 8.23). A scale factor less than 1 reduces the size and greater than 1 increases it. If you increase the scale too much the object might disappear. It hasn't gone anywhere, it's just that you are zoomed in on a single brick. Zoom back out to see the whole thing.

CAmera/TArget/Distance/POints/PAn/Zoom/TWist/CLip/
  Hide/Off/Undo/<eXit>: **Z**
Adjust zoom scale factor <1>: Use slider bar and press <**ENTER**>
CAmera/TArget/Distance/POints/PAn/Zoom/TWist/CLip/
  Hide/Off/Undo/<eXit>: **PA**
Displacement base point: Pick any point.
Second point: Move cursor around and enter ^C to cancel the pan.

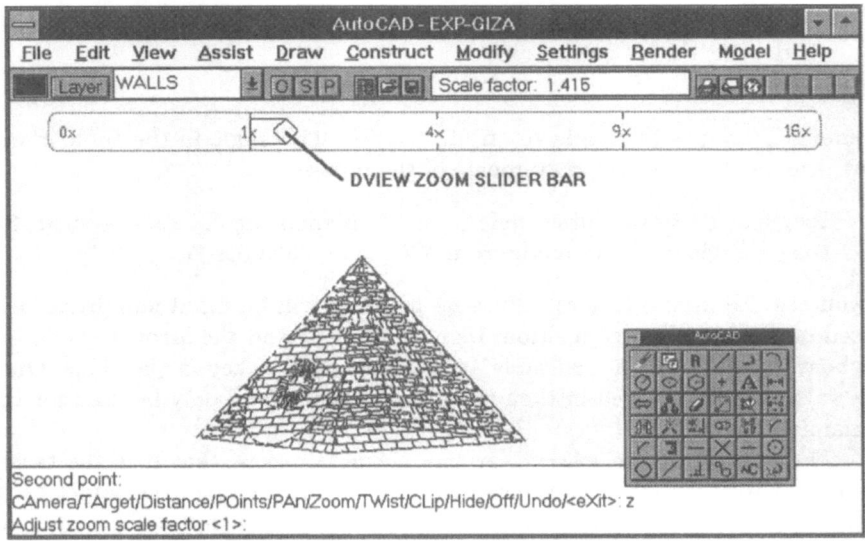

Figure 8.23 Dview zoom with slider

TWist lets you rotate the view in the plane of the screen about the target point. This has the effect of rotating the camera on the line of sight. A rubber band appears from the target to the cursor cross-hairs and shows the current angle of twist. The angle is zero when the rubber band is horizontal and to the right. The camera is upright when the twist angle is zero, upside-down when the angle is 180 degrees and on its side for 90 degrees. The twist angle is an additional setting and does not affect the camera or target positions.

> CAmera/TArget/Distance/POints/PAn/Zoom/TWist/CLip/
>   Hide/Off/Undo/<eXit>: **TW**
> New view twist <0.00>: Move cursor around and press <**ENTER**>

You can generate cut-away images with the CLip option. This allows you to specify planes in front of and behind the target to cut through the object. Nothing between the camera and the front plane will be displayed. Similarly nothing behind the back plane is shown. This can be used to eliminate unecessary foreground and background detail or to generate a cut-away view.

> CAmera/TArget/Distance/POints/PAn/Zoom/TWist/CLip/
>   Hide/Off/Undo/<eXit>: **CL**
> Back/Front/<Off>: **F**

Eye/ON/OFF/<Distance from target> <464.94>: **215**

A front distance of 464.94 places the plane at the camera in this instance while 0 would put it at the target point. Thus, 215 is between the camera and target, within the limits of the pyramid. The "Eye" option places the front plane at the camera point. This is useful for perspective views when clipping cannot be turned off. In normal dynamic viewing you can turn the front clip ON and OFF.

To get a better view of the front clip use the Hide option (Figure 8.22). This does a hidden line removal just like the **HIDE** command. Then clip a piece off the back and remove the hidden lines.

CAmera/TArget/Distance/POints/PAn/Zoom/TWist/CLip/
  Hide/Off/Undo/<eXit>: **H**
Hiding lines: done 100%
CAmera/TArget/Distance/POints/PAn/Zoom/TWist/CLip/
  Hide/Off/Undo/<eXit>: **CL**
Back/Front/<Off>: **B**
ON/OFF/<Distance from target> <−149.93>: **−300**
CAmera/TArget/Distance/POints/PAn/Zoom/TWist/CLip/
  Hide/Off/Undo/<eXit>: **H**
Hiding lines: done 100%

The minus indicates that the plane is to be behind the target. These clipping planes will remain in effect until CLip is turned off.

Getting things in perspective

To get a realistic view of objects you can generate a perspective view. When objects are in perspective the ones nearer the camera appear bigger than those further away. You can control the perspective view by choosing the "Distance" option from DVIEW. You specify the distance from the camera to the target point and AutoCAD calculates the appropriate sizes of the objects. Again, a slider bar is available to input the distance via the mouse, and the status line gives a read-out of the current slider bar position.

CAmera/TArget/Distance/POints/PAn/Zoom/TWist/CLip/
  Hide/Off/Undo/<eXit>: **D**
New camera/target distance <464.94>: **1500**

WARNING! The UCS icon should change to a perspective view of rectangular block. Most of AutoCAD's commands do not work on perspective views, so keep your eye out for this icon.

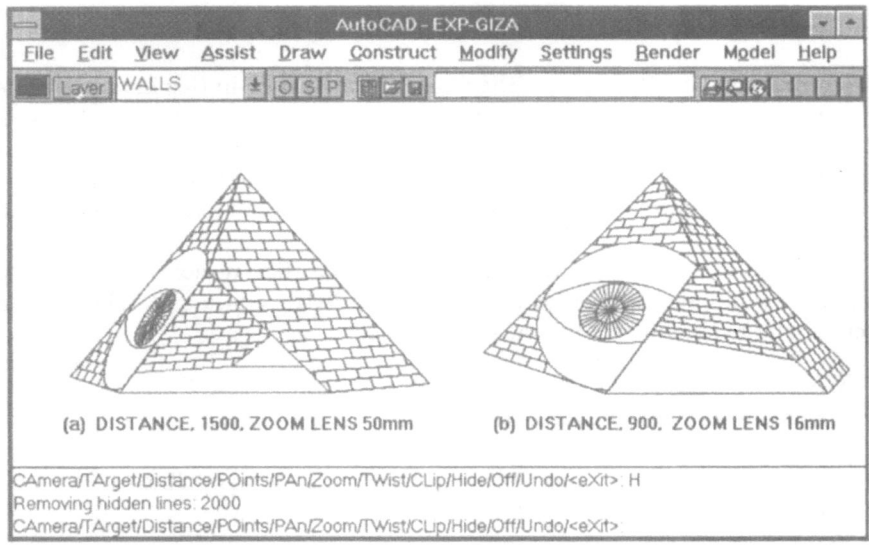

Figure 8.24 Perspective views

The zoom operates slightly differently when a perspective view is displayed. Instead of a scale factor AutoCAD asks for a camera lens length. The default is 50mm which is the standard lens focal length for most cameras. Making the lens length longer is like using a telephoto lens and magnifies the image. A shorter lens length simulates a wide angle camera lens which accentuates the perspective (Figure 8.24). The slider bar gives the zooms in multiples of the current lens length.

CAmera/TArget/Distance/POints/PAn/Zoom/TWist/CLip/
   Hide/Off/Undo/<eXit>: **Z**
Adjust lenslength <50.000mm>: **16**
To get an interesting fish-eye lens effect, shorten the perspective distance and do another hide.

CAmera/TArget/Distance/POints/PAn/Zoom/TWist/CLip/
   Hide/Off/Undo/<eXit>: **D**
New camera/target distance <1000>: **900**
CAmera/TArget/Distance/POints/PAn/Zoom/TWist/CLip/
   Hide/Off/Undo/<eXit>: **H**
Hiding lines: done 100%

To round off the DVIEW command, the "Off" option turns the perspective viewing off. "Undo" goes back to the previous view and "eXit" leaves the DVIEW command. The display retains the DVIEW settings on exit. It is dangerous to remain in a perspective view so switch it off. If there are any dviews that you wish to retain then exit the command and save them as a named view. The named view will retain all the information including any perspective setting and clipping planes.

> CAmera/TArget/Distance/POints/PAn/Zoom/TWist/CLip/
>   Hide/Off/Undo/<eXit>: **Off**
> CAmera/TArget/Distance/POints/PAn/Zoom/TWist/CLip/
>   Hide/Off/Undo/<eXit>: **X**
> Command: **QSAVE**

That concludes the AutoCAD Express stop in Egypt. It doesn't conclude the exploration of AutoCAD's third dimension. The next section covers more exciting features.

AutoCAD's 3D box of tricks

You have already used the 3DFACE command and drawn lines in 3D space. AutoCAD contains a range of facilities for creating complicated 3D objects containing many sides and faces. You can also generate smooth surfaces and 3D polylines. You will find these in the 3D sub-menu.

The 3DPOLY command allows you to create a polyline in 3D space. The PLINE command is restricted to 2D. Points can be specified in the same way as for drawing a 3D line but the use of 3DPOLY is restricted to straight line segments. The line width is zero and cannot be changed. Neither can you draw a 3D polyarc. You can use PEDIT on a 3D polyline to change any of the vertices or to fit a spline curve to the points.

Start up a new drawing with limits (0,0) to (1200,850). You will use this to draw examples of the 3D constructions. Pick **File/New** and, using ACAD as the prototype, call it **EXP-3D**.

> Command: **LIMITS**
> Reset Model space limits
> ON/OFF/<Lower left corner> <0.00,0.00>: **<ENTER>**
> Upper right corner <420.00,297.00>: **1200,850**
> Command: **ZOOM**
> All/.../Window/<Scale (X/XP)>: **A**
> Command: **3DPOLY** or Pick **Draw/Polyline/3D**
> From point: **100,100,0**   (A)
> Close/Undo/<Endpoint of line>: **@125,0,25**   (B)

Close/Undo/<Endpoint of line>: **@0,125,25**   (C)
Close/Undo/<Endpoint of line>: **@−130,0,25**   (D)
Close/Undo/<Endpoint of line>: **@0,−130,25**   (E)
Close/Undo/<Endpoint of line>: **@135,0,25**   (F)
Close/Undo/<Endpoint of line>: **@0,135,25**   (G)
Close/Undo/<Endpoint of line>: **<ENTER>**

Note that the 3D polyline cannot be assigned a width. It can be edited in other ways using PEDIT. For example make this polyline into a spiral (Figure 8.26).

Command: **PEDIT**
Select polyline: **LAST**
Close/Edit vertex/Spline curve/Decurve/Undo/eXit <X>: S
Close/Edit vertex/Spline curve/Decurve/Undo/eXit <X>: **<ENTER>**

PEDIT recognized that the object is a 3D polyline and offers only the relevant editing options. The accuracy of the fit is controlled by a system variable "SPLINESEGS". If the system variable "SPLFRAME" is non-zero then the original polyline is also shown. This gives an idea of how accurately the curve fits the points. Do this and change the view point to see the graceful spiral. Non-zero SPLFRAME also shows up any invisible 3DFACE edges.

Command: **SPLFRAME**
New value for SPLFRAME <0>: **1**
Command: **VPOINT**
Rotate/<View point>: <0.00,0.00,1.00>: **1,−1,1**

Meshes

You can build up the surface of an object by using lots of 3DFACEs but it could take a long time. A slight improvement is to use the 3DMESH command but this still requires you to input the coordinates of each vertex. If the object has a regular shape then the mesh can be generated from various control lines using either EDGESURF, REVSURF, RULESURF or TABSURF.

The best way to make a MESH is to select "mesh" from the 3D Objects dialogue box shown in Figure 8.25. You then specify the number of vertices. These are given in the length and breadth directions (M and N, respectively). First, set SPLFRAME to zero again.

Command: **SPLFRAME**
New value for SPLFRAME <1>: **0**
Pick **Draw/3D Surfaces/3D Objects...**

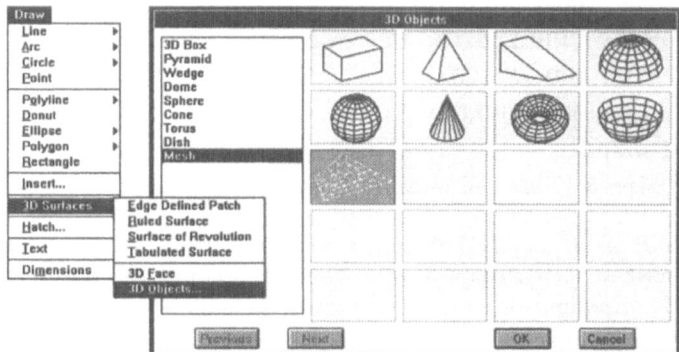

Figure 8.25 3D Objects

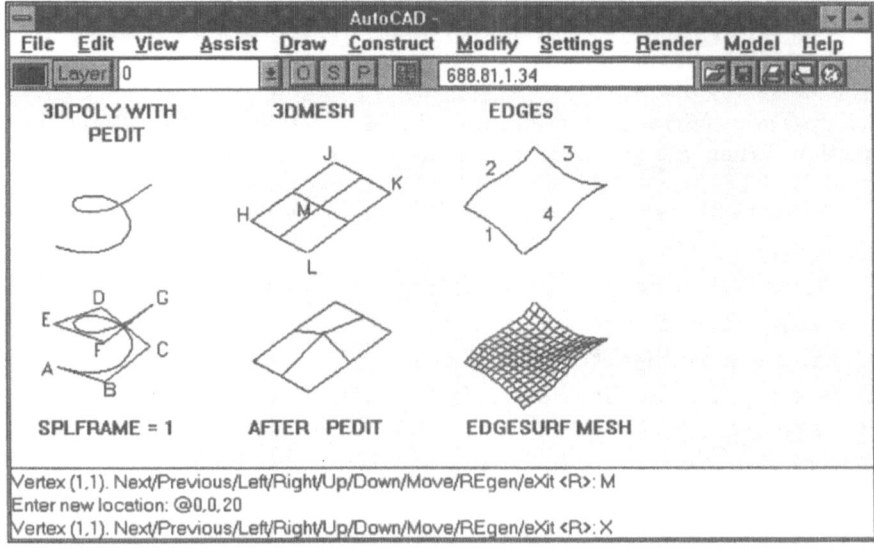

Figure 8.26 3D spiral and meshes

Pick **mesh** or pick the icon in Figure 8.25 and **OK**

Command: ai_mesh

Initializing. . . 3D Objects loaded.

First corner: **300,100,50**   (H)

Second corner: **300,250,70**   (J)

Third corner: **400,250,70**   (K)

Fourth corner: **400,100,50**   (L)

Mesh M size: **3**

Mesh N size: **3** (This will make four faces in the mesh.)

You can use PEDIT on 3DMESHes as well (Figure 8.25). The edit options are different when a mesh is selected. You can smooth the mesh, close it in either M or N directions or change individual vertices. Use PEDIT to raise the middle vertex (1,1), M, by 20 units. Do a **Zoom All** if you can't see the whole mesh.

Command: **PEDIT**

Select polyline: **LAST**

Edit vertex/Smooth surface/Desmooth/Mclose/Nclose/Undo/
   eXit <X>: **Edit**

Vertex (0,0). Next/Previous/Left/Right/Up/Down/Move/REgen/
   eXit <N>: **U**

The Up/Down refers to movements in the M direction and Left/Right the N direction. When you get to the desired vertex you can "Move" it.

Vertex (1,0). Next/Previous/Left/Right/Up/Down/Move/REgen/
   eXit <U>: **R**

Vertex (1,1). Next/Previous/Left/Right/Up/Down/Move/REgen/
   eXit <R>: **M**

Enter new location: **@0,0,20**

Vertex (1,1). Next/Previous/Left/Right/Up/Down/Move/REgen/
   eXit <R>: **X**

Edit vertex/Smooth surface/Desmooth/Mclose/Nclose/Undo/
   eXit <X>: **<ENTER>**

Note that the numbering of the vertices starts at (0,0) and so (1,1) is the second across and second up. Smoothing is only relevant when there are more than two faces in one of the directions. The command AI_MESH is a big improvement on the old 3DMESH command. It does have a silly name though.

Generated surfaces

The 3DPOLY is quite good for defining the edges of a surface. Once the edges are known, EDGESURF, RULESURF, TABSURF and REVSURF can be used to fill in the surface. TABSURF requires one edge and an extrusion direction, REVSURF needs a profile edge and an axis of revolution. RULESURF is defined by two edges, while EDGESURF is the most complicated, requiring four edge curves.

EDGESURF works by interpolating a Coons surface patch between four curves. The Coons patch is a mathematical technique using two cubic equations. The edges can be made up of lines, arcs, or open polylines and must touch at their end points. Use 3DPOLY to create four connected curves and then pick **Draw/3D Objects** and **Edge Defined Patch** from the menu bar (Figure 8.25. This executes the command EDGESURF to make a mesh like the one in Figure 8.26.

> Command: **EDGESURF**
> Select edge 1: Pick the first edge curve.
> Select edge 2: Pick the second edge curve.
> Select edge 3: Pick the third.
> Select edge 4: Pick the fourth.

This generates a polygon mesh which can be edited in the same way as the previous mesh. The vertices are numbered with the M direction along the first edge curve. The (0,0) vertex will be at the end point of the first edge nearest to the pick point used to select it. The number of faces that are created depends on the values of the two system variables, SURFTAB1 for the M direction and SURFTAB2 for N. The defaults for these are 18 each and they can be changed by typing Surftb1 or Surftb2 at the command prompt.

You must have four edges to define an EDGESURF mesh. If the shape requires only three curves then use BREAK to split one of the sides in two.

If some of the sides can be defined by straight lines or regular shapes such as arcs and circles then the commands TABSURF, RULESURF or REVSURF may be more appropriate (see Figure 8.27).

The Tabulated Surface or TABSURF is good for extruding objects in 3D space. It gives an effect similar to setting an entity THICKNESS but is more general. To generate a TABulated SURFace you require some object, called path curve in AutoCAD, to extrude and a line defining the direction of extrusion. In descriptive geometry jargon you need a *directrix* object and a *generatrix* vector. To make a leaning tower draw a circle in the WCS plan and a 3D line.

> Command: **CIRCLE**
> 3P/2P/TTR/<Center point>: **500,300,0**

Diameter/<Radius>: **50**
Command: **LINE**
From point: **570,300,0**
To point: **@0,40,100**
To point: **<ENTER>**
Command: **TABSURF** or Pick **Tabulated Surface** from the menu.
Select path curve: Pick the circle.
Select direction vector: Pick the line.

This generates an open ended leaning tower. It is not cylindrical though, since the direction vector is not perpendicular to the plane of the circle. The extrusion direction depends on the point order used when the line was drawn.

A Ruled Surface is more general than the tabulated surface. You specify two boundary edges and RULESURF joins them together with straight lines to form a polygon mesh. You can use open or closed 2D and 3D polylines, circles, arcs, lines and points. However you cannot mix a closed object such as a circle with an open object such as a line. Points can be used with either open or closed paths. Make a surface from an arc to a line. Remember, ARCS, like circles are drawn in the plan of the current UCS.

Command: **ARC**
Center/<Start point>: **800,300,0**
Center/End/<Second point>: **C**
Center: **750,300**   (You must give a 2D point.)
Angle/Length of chord/<End point>: **A**
Included angle: **270**
Command: **LINE**
From point: **800,500,0**
To point: **@−220,0,30**
To point: **<ENTER>**
Command: **RULESURF** or pick **Ruled Surface** from the menu.
Select first defining curve: Pick the arc near the start point.
Select second defining curve: Pick the line near (800,500,0).

The two end points nearest the places where the curves are picked define the starting vertices of the mesh. If you pick one of the curves near the wrong end the surface will be twisted. If this happens use the UNDO command and try again, picking the points at the correct ends of the curves. The number of divisions for both TABSURF and RULESURF is determined by the SURFTAB1 system variable.

The final surface generator in the 3D box of tricks is REVSURF. This produces a surface of revolution from a definition path and an axis to rotate it around. Common surfaces of revolution include wine goblets, spheres, torus shapes and power station cooling towers. REVSURF allows you to make either

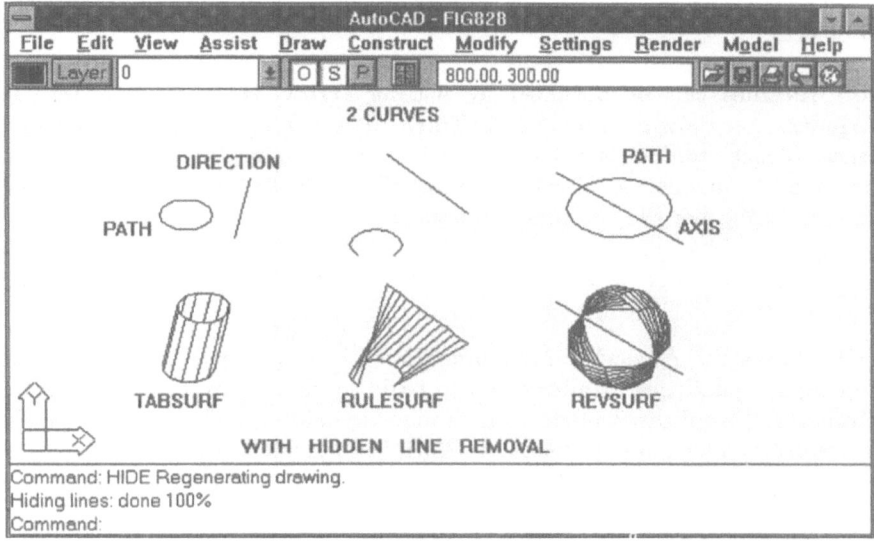

Figure 8.27 3D generated surfaces

closed surfaces or open ones by controlling the angle of rotation. To make a part sphere draw a circle with an axis along one diameter.

> Command: **CIRCLE**
> 3P/2P/TTR/<Center point>: **1000,300,0**
> Diameter/<Radius>: **100**
> Command: **LINE**
> From point: **860,300,0**
> To point: **@280,0**
> To point: **<ENTER>**
> Command: **REVSURF** or pick **Surface of Revolution** from the menu.
> Select path curve: Pick the circle.
> Select axis of revolution: Pick the line.
> Start angle <0>: **<ENTER>**
> Included angle (+ccw −cw) <Full circle>: **90**

The positive direction of the axis depends on where the line defining the axis is picked. Angles are positive in the anti-clockwise direction as you look from the picked point to the furthest end point.

3D objects

More routines can be obtained by picking **Draw/3D Surfaces** and **3D Objects...** as shown in Figure 8.25. These allow you to easily produce spheres, toruses, cones etc. The number of divisions in the sphere, hemi-sphere, torus and cone is controlled by SURFTAB1 and SURFTAB2. The larger these numbers, the smoother the surfaces will appear.

AME

AME stands for Advanced Modelling Extension and gives AutoCAD solid modelling capabilities. It allows you to build up models from basic shapes or primitives. Complicated shapes can be made by adding, subtracting or getting the intersection or union of solids. As AME is not a standard part of AutoCAD but an add-on a full description is beyond the scope of this book. Indeed, it is itself the subject of a number of books.

Summary

This chapter has covered the use of AutoCAD in isometric projection, 2.5D and full 3D. You have also encountered the display features of user coordinate systems and view points. The UCS and VPORT commands can be used in 2D drafting as well as 3D. Dynamic viewing is a powerful aid to visualizing 3D spatial relationships.

The UCS facility is the most important tool in 3D computer aided drafting. Many objects are 2D (eg arcs, circles) and can only be drawn in plan. To draw a sloping circle you have to create a coordinate system so that the plane of the circle is the same as the plane of the UCS.

Using 3D CAD involves an extra level of difficulty above 2D drafting and requires much more discipline. You must keep track of where objects are and also what coordinate system and view point is being used. Vigilance helps to prevent troublesome errors caused by optical illusions.

You should now be able to:

Draw lines and circles in isometric projection.
Create objects with different elevations and thicknesses.
Use hidden line removal.
Set up multiple viewports.
View 2.5D and 3D objects from different view points.
Define new coordinate systems in 3D space.
Draw objects in 3D space.

Run dynamic visualisations and create perspective views.
Generate 3D surfaces.

Chapter 9 THE HARDCOPY – PRINTING AND PLOTTING

General

The main purpose of producing drawings with AutoCAD is to communicate graphical information. Even at this late stage in the twentieth century, the primary medium for such communication is with pictures on paper. Paper drawings are very user friendly in that they are easy to read and transport and also provide a useful framework for rough work and checking. With this in mind the current chapter is devoted to methods of translating the digital information in the AutoCAD drawing into black marks on paper.

A new order of activity is involved in producing plots and prints. That is you have to control another piece of equipment, be it printer, plotter or both. According to Murphy's Law this extra complexity inevitably leads to more things that will go wrong. To avoid the heartache and frustration associated with peripheral blues, stay calm and follow the guidelines laid out below.

Unless you have access to very expensive plotting equipment the generation of a hardcopy of a drawing takes time. The cheaper your plotter the longer it will take. In general you will not want to have to reproduce plots too often and so the aim is to get it right first time, if at all possible.

You will produce copies of some drawings from earlier chapters. Make sure that the drawing files BALLOON.DWG, from Chapter 4, GLAND.DWG, from Chapter 7, and EXP-GIZA from Chapter 8 are handy. You will also create a standard title block in AutoCAD's paper space and generate a multiple view plot of the all seeing pyramid. If you don't have these files you can improvise with some other simple drawings. Avoid plotting bigger drawings until you have more experience. It is assumed that AutoCAD has been configured correctly for your printer and plotter. If you are not sure if this has been done refer to Appendix A or the *AutoCAD Interface, Installation and Performance Guide*.

Printing and plotting

Printers

Drawings can be output on various types of printers, from humble 9-pin dot matrix to whizz-bang color laser printers. Daisywheel printers do not support graphics and so are no good for AutoCAD. The quality of the printer plot depends on the resolution capability of the device. The printer produces the picture by converting all the graphics into a series of dots which are then inscribed on the paper. The crucial statistic for assessing resolution is the number of *dots per inch* (dpi) capability of the printer. The higher the dpi, the better the picture quality. This is most noticeable when plotting arcs and other curved entities and shallow sloping lines.

With laser printers, you must also check the maximum area that can be plotted at the highest resolution. A sneaky trick by some printer manufacturers was to advertise a high resolution of say 300dpi but conceal in the small print that this could only be achieved for drawings with a total area of less than 6 square inches. Most laser printers and inkjet printers being sold now will allow at least 300dpi for a full page.

Plotters

The distinction between some plotters and printers is getting difficult to see. There are now A0 plotters based on inkjet and laser printer technology. There are also many varieties of pen plotter which allow cheap multi colored output. With proper selection of pens and plotting medium you will get great results. However, compared with inkjet and electrostatic plotters they are slow.

Note that if you use different printers and plotters from time to time, you will have to configure AutoCAD for each one. AutoCAD can store multiple output device configurations. See Appendix A for details.

HAZARD WARNING! As a general precaution, before issuing any print or plot instructions from within the AutoCAD editor you should save the drawing. This is good CAD practice as a malfunction of the peripheral can cause the computer to hang up. If you have to reboot the system you will lose everything drawn since the last SAVE command.

Plotting the GLAND drawing

For the first plot, you need to start up AutoCAD and OPEN the Gland drawing from Chapter 7. To do this select **File/Open** from the menu bar. Then click

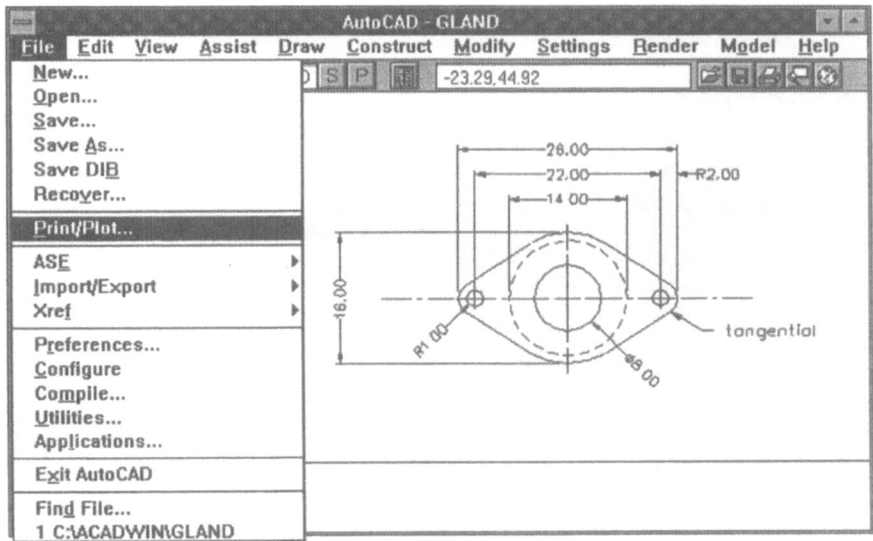

Figure 9.1 Plotting the gland

on **gland.dwg** from the list of file names. When the file has been loaded, pick **Print/Plot.**.. from the **File** menu (Figure 9.1).

The first thing to do when the Plot Configuration dialogue box appears is to check which output device you are going to use. Pick the button marked **Device and Default Selection** in the top left part of the dialogue screen (Figure 9.2). This will give a list of the available printers and plotters. Pick the device of your choice. In this example, I have chosen my A1 pen plotter. Then pick **OK**. If your device does not appear in the list, then you will have to configure AutoCAD as described in Appendix A.

Altering the plot parameters

Remember, the limits were set to (0,0) and (65,45) in Chapter 7. However, as the actual drawing doesn't quite fill the limits we will plot the "Extents" of the drawing. Specifying **Extents** in the Additional Parameters section of the plot dialogue box shown in Figure 9.2 causes AutoCAD to calculate a window just big enough to fit in all the drawings entities. The main alternatives to choosing extents are using the drawing limits or the current display. The Window button behaves as with the ZOOM command and allows you to plot part of the drawing.

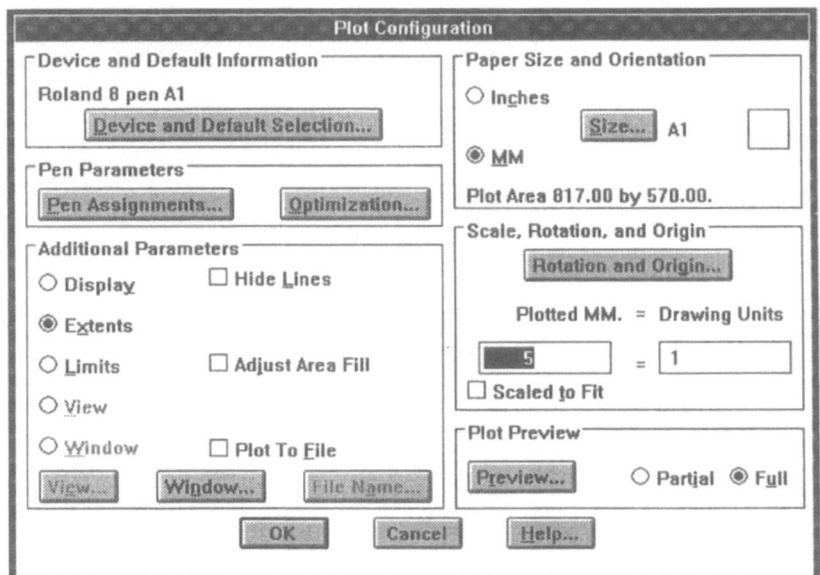

Figure 9.2 Plot dialogue box

The "Hide Lines" is only applicable for 3D drawings and is dealt with later. The "Adjust Area Fill" is used when the edges of the solids and polylines must be positioned very accurately. If the area is adjusted, then the pen strokes are moved in by one half the pen width at the boundaries. It saves AutoCAD some calculation time if the adjustment is not requested.

In some cases, it is desirable to generate plot files which can be sent to the printer later. For example, if you had a large number of drawings to print or if the printer was not available, you could still generate the plot files, store them on disk and print them later. Examples of this are treated later.

Even though my plotter can handle A1 sized paper (840mm by 594mm) the Gland drawing doesn't quite need it all. This plot is to be produced on an A4 sheet in landscape ie 297mm by 210mm. This can be set by picking the **Size** button near the top right of the dialogue screen. In this example, I have also set the units to millimeters. Picking the Size button gives yet another dialogue box with all the available paper sizes for the device (Figure 9.3). Note that the A4 size given by AutoCAD includes an allowance for the plotter's margin. Most devices cannot go right to the edge of the paper. In general the margins allowed for by AutoCAD are incorrect particularly if your plotter is running some form of emulation. Therefore you will most likely need to input your own paper size. Here, the full A4 size of 297 by 210 has been input as the

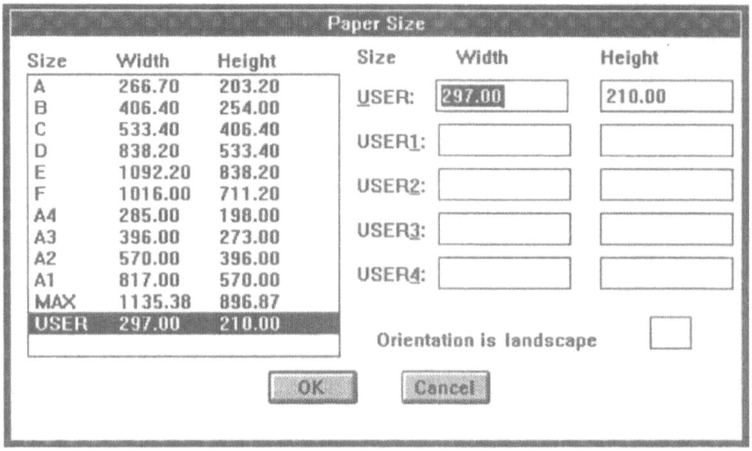

Size	Width	Height	Size	Width	Height
A	266.70	203.20	USER:	297.00	210.00
B	406.40	254.00			
C	533.40	406.40	USER1:		
D	838.20	533.40			
E	1092.20	838.20	USER2:		
F	1016.00	711.20			
A4	285.00	198.00	USER3:		
A3	396.00	273.00			
A2	570.00	396.00	USER4:		
A1	817.00	570.00			
MAX	1135.38	896.87			
USER	297.00	210.00	Orientation is landscape		

Paper Size

OK   Cancel

Figure 9.3 Paper size

USER size. This then gets added to the list of sizes on the left when you press <**ENTER**>. You pick the appropriate size from this list and then pick **OK** to go back to the previous dialogue box.

Scales and Preview

After you have chosen the paper size, AutoCAD calculates a scale factor to fit. In this case it comes up with a silly scale of "297 plotted mm = 57.9 drawing units. Engineers and designers are used to reading drawings at specific scales (2:1, 1:1, 1:5 etc). Faced with a scale of say 5.129:1, difficulties arise. Thus, here scale will be prescribed by the user. Click on the **Scaled to Fit** option to remove the "X". Then input **5** for the plotted mm and **1** drawing unit as shown in Figure 9.2. Thus if each drawing unit represented 1mm in real life the plot will be 5:1 ie five times bigger than reality.

Before executing the plot it's best to get a preview of what the output will look like. The Plot Preview button is towards the lower right of the dialogue box (Figure 9.2). There are two options, Partial or Full. A partial preview will show only the limits of the plot in relation to the paper. This is useful for checking that the orientation and origin are sensible. However, if parts of your drawing exceed the paper size you will not know from this. A full preview will show exactly what the plot will look like. Pick the **Full** button and then pick **Preview.** Your display should then resemble Figure 9.4. If the "Plot Preview" control window obscures part of the image you can move it by picking and

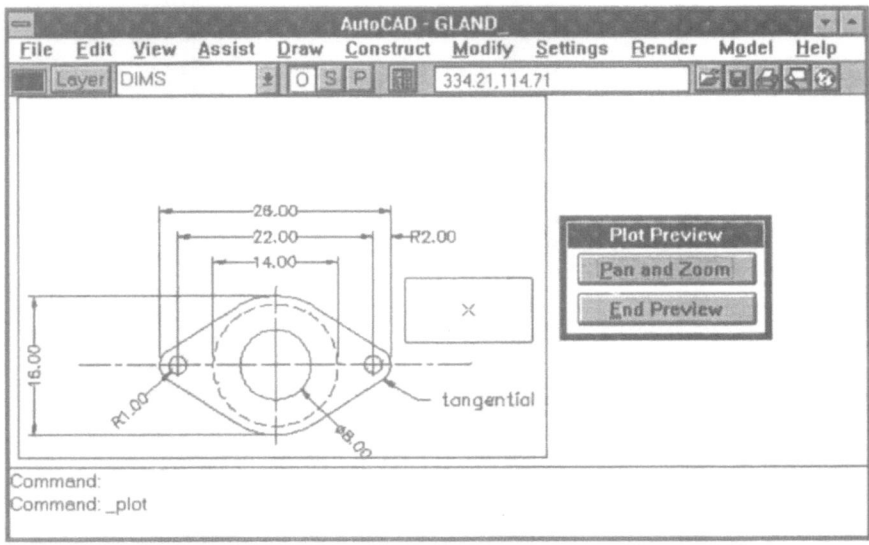

Figure 9.4 Previewing the plot

dragging the title bar. The Pan and Zoom work like Zoom/Dynamic. Pick the **Pan and Zoom** button. This gives the rectangle with an "X" in the middle. This can be dragged around by moving the mouse. When you reach the part you wish to magnify press the mouse button. The "X" changes to an arrow at the right-hand edge of the rectangle. Moving the arrow changes the size of the window. Press **<ENTER>** to show the magnified view. Pick the **Zoom Previous** button to go back. Pick **End Preview** to get back to the plot configuration dialogue box.

On the preview shown in Figure 9.4, the diameter text at the bottom is a bit too close to the edge. As there is plenty of space above the gland we can shift the plot origin up a bit. Pick the **Rotation and Origin** button from the Plot Configuration dialogue box. Then change the **Y Origin** to **30** as shown in Figure 9.5. The other possible options are to shift the X origin or to rotate the plot. When printing on laser printers it is usually necessary to rotate the plot by 90 degrees. This is because printers treat paper in portrait mode ie taller than the width, while drawings are usually in landscape mode, ie wider than they are tall. In this example, no rotation is required so pick **OK**. Check the preview once more. If all is well, then pick **OK** in the Plot Configuration dialogue box.

AutoCAD will then echo:

Command: _plot

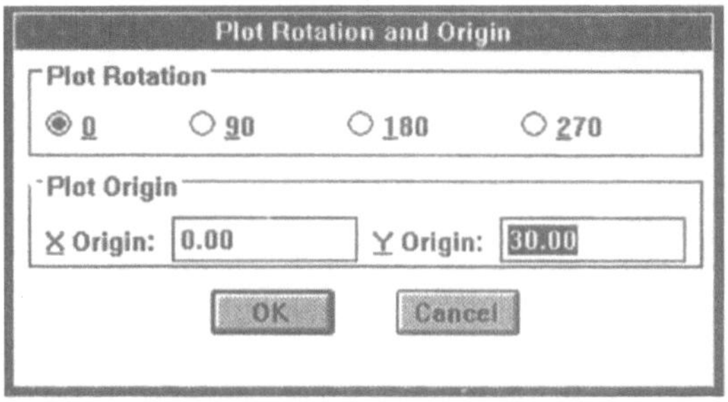

Figure 9.5 Shifting the origin

Effective plotting area: 289.49 wide by 152.56 high
Position paper in plotter
Press RETURN to continue or S to Stop for hardware setup

AutoCAD then pauses to allow you to get the plotter and paper ready. If you press <**ENTER**> at this prompt AutoCAD will send a plotter reset function and commence plotting. This "reset" clears the plotter memory buffer and may also clear any special plotter settings. Some plotters allow you to key in specific pen speeds and pressures using a control panel. To avoid the "reset" disturbing these settings you can press S at the prompt. This sends the "reset" but then pauses to allow the setup. When that is complete you press <**ENTER**> to begin the plot. AutoCAD then keeps you informed of the progress of the plot:

Regeneration done xx%

When this is 100% you will get the "Plot complete" message.

Saving a VIEW

In this part of the exercise, you will generate a number of multicolored plots of the Balloon drawing from Chapter 4. It is often necessary to produce a number of different plots of various parts of a drawing. You might want a large scale view of some detail as well as a general layout. You can explore the drawing using ZOOM, PAN and VPOINT or DVIEW. Once you are happy with the display it can be stored as an AutoCAD VIEW. Such views can be quickly retrieved for redisplay or plotting. This saves you from having to remember what ZOOMs and other operations were used to get the desired effect. In this section you will create two VIEWs of the BALLOON drawing.

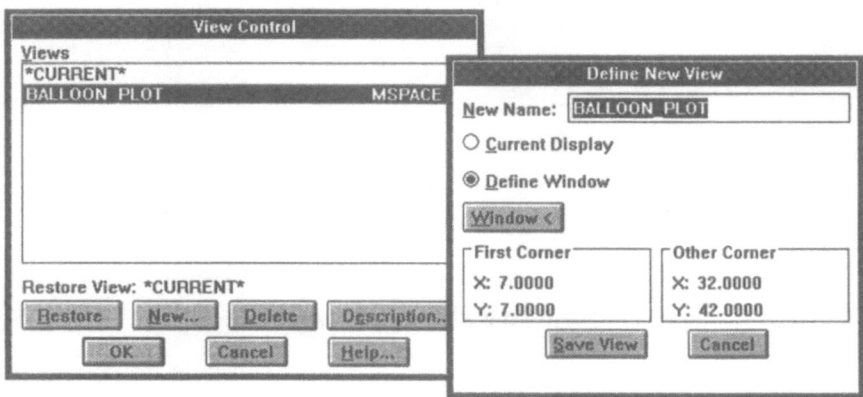

Figure 9.6 View creation

To open the drawing file pick **File/Open** followed by **balloon.dwg** from
the file list. Named views are created by picking **View/Set View** followed
by **Named View. . . .** In the View Control screen (Figure 9.6) pick **New**. This
allows you to define a new view. Give the name as **BALLOON_PLOT** and
pick the circle beside "Define Window". Then pick the **Window** button. This
returns you to the graphics screen to pick a window or type the coordinates.

Command: _ddview
First corner: **7,7**
Other corner: **32,42**

Finally, pick **Save View** to add it to the list shown in the view control box.

This allows you to save a rectangular section of the display. Of the other
options, "Description" gives details of the settings associated with the view;
"Delete" allows you to delete a stored view (but not the one named "*CUR-
RENT*"; "Restore" causes a named view to be displayed on the graphics
screen. In model space, if you have more than one viewport on the screen only
the active one can be stored as a VIEW. The rules for view names are the
same as for layers, up to 31 numbers or letters but no spaces or full stops. The
"MSPACE" after the view name indicates that is was created in model space.

Note that this is a tall, thin VIEW. When it is displayed on the screen,
parts of the drawing to the left and right of the view's window may be displayed. However, when it is plotted only that portion within the window defined by (7,7) to (32,42) will be drawn. To see the view (Figure 9.7), pick the
line:

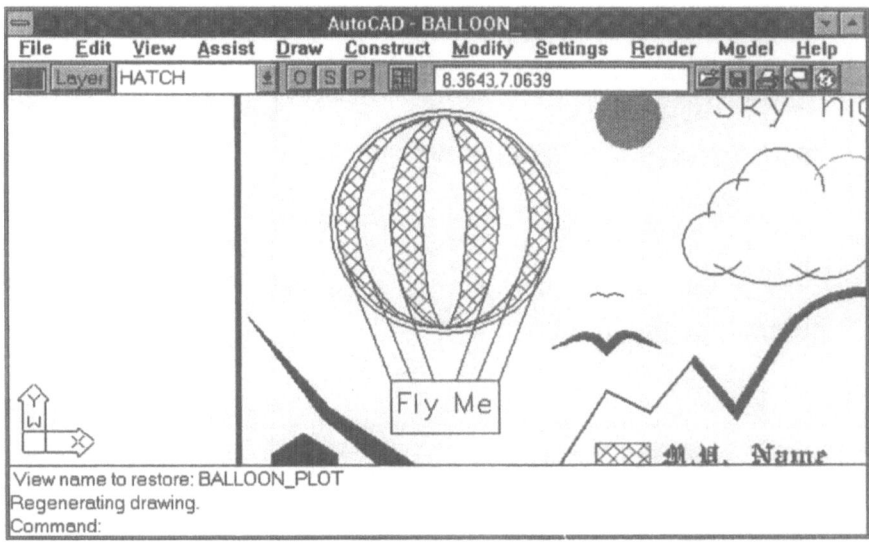

Figure 9.7 BALLOON_PLOT view

BALLOON_PLOT   MSPACE

followed by the **Restore** button and the **OK**.

To generate the second view that is plotted in Figure 9.10, ZOOM in to the lower right corner and use the "Current Display" option in the Define New View dialogue box.

Command: **ZOOM**

All/.../Window/<Scale(X/XP)>: **W**

First corner: **30,0**

Other corner: **65,25**

Command: **DDVIEW** or pick View/Set View/Named View...

When the View Control dialogue box (Figure 9.6) appears pick **New...**, give the new name as **MOUNTAIN_PLOT**, pick the **Current Display** button followed by the **Save View** button. To exit the View Control pick **OK**. You now have two suitable views for plotting.

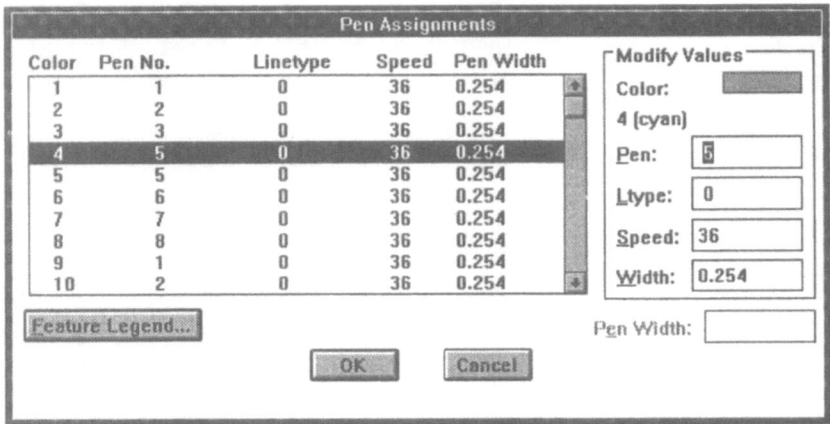

Figure 9.8 Setting pens

Multi-colored plotting

With most plotting devices you can have more than one pen and so vary the colors or line thicknesses. You can also vary the pen speed on some plotters to get darker or lighter lines.

Remember to save the drawing before plotting. Use the quick save command, picking **File/Save**. To generate a plot of the mountain view pick **File/Print/Plot...** from the menu bar. Check the pen settings by picking the **Pen Assignments** button from the Plot Configuration screen. The dialogue box shown in Figure 9.8 should then appear.

The first column indicates the AutoCAD color number, 1=red, 7=black etc. Next comes the pen numbers. These will depend on how you have loaded you plotter. Making modifications is similar to using the Layer Control dialogue box. First you select the line or lines you wish to modify from the left hand list and then use the options on the right. Note that AutoCAD has 256 colors but there are only 8 on this plotter. The higher numbered colors are mapped to pens in sequence using multiples of eight. Thus color number 9 is mapped onto pen 1. In Figure 9.8 color number 4, cyan, has been reset to use pen number 5. This is done by picking the line:

4   4   0   36   0.254

Then using the "Modify Values" area on the right, move to the **Pen:** field and key in **5**. You can also change the plotter linetype, pen speed and width setting. The linetypes here are not AutoCAD linetypes but plotter ones. It

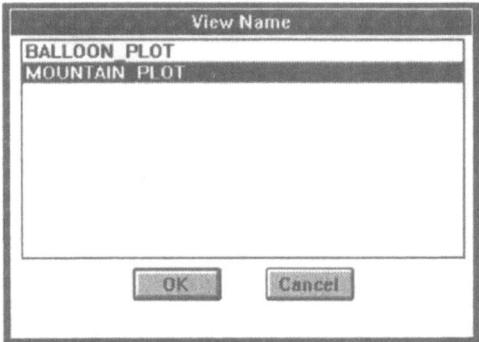

Figure 9.9 Selecting a view

is not recommended to mix AutoCAD and plotter linetypes so it is best to leave Ltype at 0. The pen speed option would be useful if it worked. With my Roland, emulating a Hewlett Packard device it doesn't. Thus it is safest to set the pen speeds using the plotter's own console. Some pens operate better at slower speeds eg drafting pens. Finally the width can be important. AutoCAD calculates the number of passes of the pen that are required to fill in a thick polyline. If the specified width is larger than the actual pen width solid objects will appear striped rather than filled. The thicknesses in Figure 9.8 are in mm. If you change the pen thickness of say pen number 1 then the change is applied to all other colors using that pen. Thus color 9, 17, etc would also have the new width.

Plotting a view

Having made all the modifications you require to the pen assignments, pick **OK** to get back to the Plot Configuration screen. Now pick the **View...** button in the lower left corner. This gives a window with all the views defined in the drawing (Figure 9.9). Pick **MOUNTAIN_PLOT** followed by **OK**.

Postscript output

An increasing number of printers and electrostatic plotters accept the Postscript command language by Adobe. Postscript is a powerful plotting language. It is hardware-independent and fast becoming the industry standard for plotting graphics and desk-top publishing. When using a Postscript device with AutoCAD you should use the configuration for "Postscript Laser Printers". The processing time for translating the drawing into Postscript commands can be

Figure 9.10 Mountain plot output

quite long. The printer may "Time out" and disconnect if it has to wait too long between commands. If this is a problem, you are recommended to send the plots to a file and then use the DOS "COPY" command. The chief advantage of using Postscript is that your plot file can be output on any make plotter that supports the language. Professional printing houses use Postscript photo plotters capable of up to 4000dpi. This means that you can plot lines as fine as 1/4000 inch! Most non-professional devices allow between 300 and 1200dpi.

Plotting to a file

There are many reasons why you might want to output a drawing to a plot-file. The plotter might not be available or you might have a lot of plots to generate. In either case you can still create your plots on disk. These plotfiles can be sent to the plotter from DOS or by using a DOS ".BAT" file or using Windows' Print Manager . Thus the AutoCAD workstation doesn't have to be tied up in lengthy plotter–computer interactions. Any workstation with DOS can be used to get the actual hardcopy. Plot files can be queued and spooled in networks. This frees the AutoCAD workstation for more productive use.

Some laser and inkjet printers can emulate plotters. This normally requires special emulation software to be run outside of AutoCAD. To generate the plot you will probably have to have the .PLT files ready for input to the printer manufacturer's software.

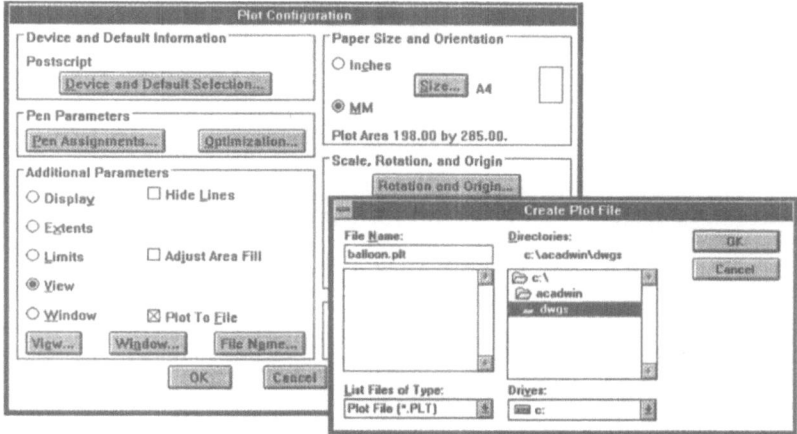

Figure 9.11 Plot to file configuration

The AutoCAD procedure is exactly the same as that already described except for the option **Plot to File** being selected from the Plot Configuration dialogue box (Figure 9.11). The default file name which is the same as the drawing but with the extension ".PLT" can be selected. However, if you have a number of plots it is best to devise your own identification system (eg LAYOUT.PLT, DETAIL1.PLT etc). Pick the **File Name...** button to bring up the Create Plot File screen also shown in Figure 9.11. This is the same format as other file creation screens. The file name can be any allowable DOS name, including the directory path. The file type ".PLT" is automatically added.

The plot shown in Figure 9.12 was generated using a Postscript laser printer. Thus the device for the plot file must be "Postscript". In the pen assignments, a width of 0.15mm was set for all colours. Anything thicker gives rather chunky looking plots. The **View** button was picked and **BALLOON_PLOT** selected. The balloon view is tall and thin. To get the biggest plot, you should use the paper in "portrait" mode. With laser printers the standard A4 size is adequate. This plot is scaled to fit the page.

Note that Figure 9.12 only includes those objects that were inside the view definition window. If you restore the view on the screen, some items to the left and right of the window are also displayed. This is because AutoCAD always tries to fill the available screen area.

As the computer doesn't have to wait for the peripheral, you should get the "Plot complete" message a lot quicker. To send the file to the plotter, you will either have to SHELL or exit from AutoCAD or switch to the Windows Program Manager and then pick the MS DOS prompt. When you are at the

Figure 9.12 Balloon Postscript plot

DOS prompt you can issue the COPY command. If your Postscript device is connected to the Parallel port, you can COPY the file to that output device.

   C:\ACADWIN> **COPY BALLOON.PLT LPT1**

If your output device is connected to the Serial port, you will have to redirect your plotter output from COM1. This can be achieved with the MODE command. This command can also be used to set the correct communications parameters for your plotter.

   C:\ACADWIN> **MODE LPT1=COM1**
   C:\ACADWIN> **COPY BALLOON.PLT PRN**

If the plotter doesn't operate properly, you may have to set the communications parameters using MODE once more. Check your plotter's documentation or the *AutoCAD Interface Installation and Performance Guide* for the correct settings. If you have emulation or plot spooling software, you should follow the supplier's instructions.

Always remember to match the AutoCAD configuration to the output device. When you are generating plot files for various plotters, you must select the appropriate Device and Default Information each time.

Plotting multiple viewports in paper space

A simple title block for EXP-GIZA

Before embarking on the exercise on paper space and the MVIEW command, let's create a title block and margin in preparation for the final plot. This small

drawing will be incorporated in the final plot of the pyramid using AutoCAD's
XREF facility. In this short section we will make a new drawing of the margins
and a title block for a standard A3 plot. Start AutoCAD if it is not still
active and create a new drawing, picking **File/New**, with the name **Title-
A3**. Accept "ACAD" as the prototype.

As with any other drawing, the first steps are to set up the limits and
create any layers that will be needed. The limits for my A3 plotter are (0,0)
to (420,297). Some plotters cannot plot up to the edge of the page and so you
might have to adjust the limits accordingly.

```
Command: LIMITS
Reset Model space limits:
ON/OFF/<Lower left corner> <0.00,0.00>: <ENTER>
Upper right corner <default>: 420,297
Command: LIMITS
Reset Model space limits:
ON/OFF/<Lower left corner> <0.00,0.00>: ON
Command: ZOOM
All/.../Window/<Scale(X/XP)>: A
Command: LAYER
?/Make/Set/...: M
New current layer <0>: MARGIN
?/Make/.../Color/...: C
Color: YELLOW or any color
Layer name(s) for color 2 (yellow) <MARGIN>: <ENTER>
?/Make/Set/...: <ENTER>
```

Now we can proceed with drawing the margins and title block shown in Figure 9.13. The title block includes some text.

```
Command: LINE
From point: 5,5
To point: @410,0
To point: @0,290
To point: @-410,0
To point: CLOSE
Command: <ENTER>
LINE From point: 315,5
To point: @65<90
To point: @100,0
To point: <ENTER>
```

Now add the internal lines of the title block.

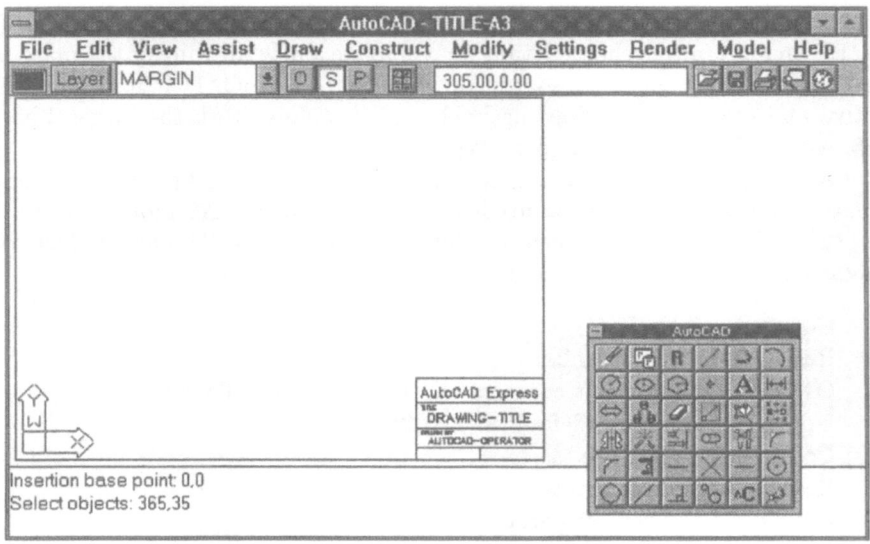

Figure 9.13 Margins and title box

Command: <**ENTER**>
LINE From point: **315,50**
To point: **@100,0**
To point: <**ENTER**>
Command: <**ENTER**>
LINE From point: **315,30**
To point: **@100,0**
To point: <**ENTER**>
Command: <**ENTER**>
LINE From point: **315,15**
To point: **@100,0**
To point: <**ENTER**>
Command: <**ENTER**>
LINE From point: **365,15**
To point: **@0,-10**
To point: <**ENTER**>

The next task is to add in the text and attributes. First make sure that the text font is SIMPLEX. Other fonts may not fit neatly in boxes with the text heights given below.

Command: **STYLE**

Text style name (or ?) <STANDARD> <**ENTER**>
New style.

Then pick **simplex.shx** from the Select Font File screen.

Height <0.00> <**ENTER**>
Width factor <1.00> <**ENTER**>
Obliquing angle <0.00> <**ENTER**>
Backwards? <N> <**ENTER**>
Upside-down? <N> <**ENTER**>
Vertical? <N> <**ENTER**>
SIMPLEX is now the current text style.
Regenerating drawing.
Command: **TEXT**
Justify/Style/<Start point>: **320,45**
Height <default>: **3**
Rotation angle <0.00>: <**ENTER**>
Text: **TITLE**
Command: <**ENTER**>
TEXT Justify/Style/<Start point>: **320,25**
Height <3.00>: <**ENTER**>
Rotation angle <0.00>: <**ENTER**>
Text: **DRAWN BY**
Command: <**ENTER**>
TEXT Justify/Style/<Start point>: **J**
Align/Fit/Center/...: **C**
Center point: **365,55**
Height <3.00>: **7**
Rotation angle <0.00>: <**ENTER**>
Text: **AutoCAD Express**

The attributes will be used to input the drawing title and CAD operator's
name. Each of the attributes must be visible and positioned using centred
text. Define the drawing title first.

Command: **ATTDEF** or pick **Draw/Text/Attributes/Define**
Attribute modes – Invisible:N Constant:N Verify:N Preset:N
Enter (ICVP) to change, RETURN when done: <**ENTER**>
Attribute tag: **DRAWING-TITLE**
Attribute prompt: **Enter the drawing title:**
Default attribute value: **No title**
Justify/Style/<Start point>: **C**
Center point: **365,35**
Height <3.00>: **7**

Rotation angle <0.00>: <ENTER>

Now define the AutoCAD operator.

> Command: **ATTDEF**
> Attribute modes – Invisible:N Constant:N Verify:N Preset:N
> Enter (ICVP) to change, RETURN when done: <**ENTER**>
> Attribute tag: **AutoCAD-Operator**
> Attribute prompt: **Enter your name:**
> Default attribute value: **Anonymous**
> Justify/Style/<Start point>: **C**
> Center point: **365,17.5**
> Height <7.00>: **5**
> Rotation angle <0.00>: <**ENTER**>

These two attributes are now grouped to form a block. Whenever the block is inserted, you will be prompted to input the drawing title and your name.

> Command: **BLOCK**
> Block name (or ?): **TTEXT**
> Insertion base point: **0,0**
> Select objects: **365,35**
> 1 selected, 1 found
> Select object: **365,17.5**
> 1 selected, 1 found
> Select object: <**ENTER**>

Now save the drawing by picking **File/Save**. The drawing is now ready to be used to generate plots. When the drawing is saved, it can be incorporated into any other AutoCAD file by either INSERTing it as a block or referencing it with the XREF command.

Paper space

You have already had a brief introduction to AutoCAD's concept of paper space in the last chapter. One of the significant features of paper space is the ability to output a number of different views of an object on one plot.

By default, drawings are produced and viewed in "model space". That is, all the coordinates have been input relative to the world in which the object was drawn. When it comes to plotting, you could choose an origin position on the paper and also a rotation angle, hidden line removal, scale etc.

Since Release 11 AutoCAD allows you to create a virtual page for plotted output. You do this by switching to "paper space". In paper space you can assign a page size, A1 for example, and a number of viewports. Different views of the object, say plan, elevation and perspective can be set up in each viewport and appropriate scales can be assigned. When all the views are correct the paper space can be plotted at a scale of 1:1.

The paper space was then filled with three viewports as shown in Figure 9.14. Note, a margin of 10mm was included when positioning these viewports. These viewports, created by the MVIEW command are themselves AutoCAD entities. Thus, they belong to specific layers etc. They can be moved, copied and erased like other entities. Furthermore, they can overlap and even control the visibility of layers and hidden lines within any view. In Chapter 8 the three viewports were created on layer 0.

Remember, in order to use AutoCAD's paper space you have to set the value of the **TILEMODE** variable to 0.

Manipulating metaviews

In this section you will generate a standard engineering type drawing of the pyramid constructed in Chapter 8. If you haven't got a copy of the drawing file, EXP-GIZA, you could improvise with any 3D object or just draw the outline of the pyramid as explained in Chapter 8. Start a new drawing using EXP-GIZA as the prototype. Pick **File/New**, if you haven't already saved the TITLE-A3 drawing do so now. Give the prototype name as **EXP−GIZA** and the new drawing name as **GIZAPLOT**. This procedure ensures that if anything goes wrong with the current exercise then at least we still have the original drawing.

The display will first resemble Figure 8.24, depending on how far you got with using DVIEW in the last chapter. To get to the situation shown in Figure 9.14 just set TILEMODE to 0. Pick **View/Tilemode** to do this. Then pick the Paper space button or type the command, **PSPACE**.

> Command: **TILEMODE**
> New value for TILEMODE <1>: **0**
> Command: **PSPACE**

In Chapter 8 when you entered paper space for the first time in a drawing you had to set the paper limits. In this case the limits were set to A3 dimensions (420mmx297mm). For the plotting exercise you need to create a new layer for the viewport entities.

> Command: **LAYER**
> ?/Make/Set/...: **M**
> New current layer <WALLS>: **PSVP**
> ?/Make/Set/...: **C**
> Color: **CYAN**

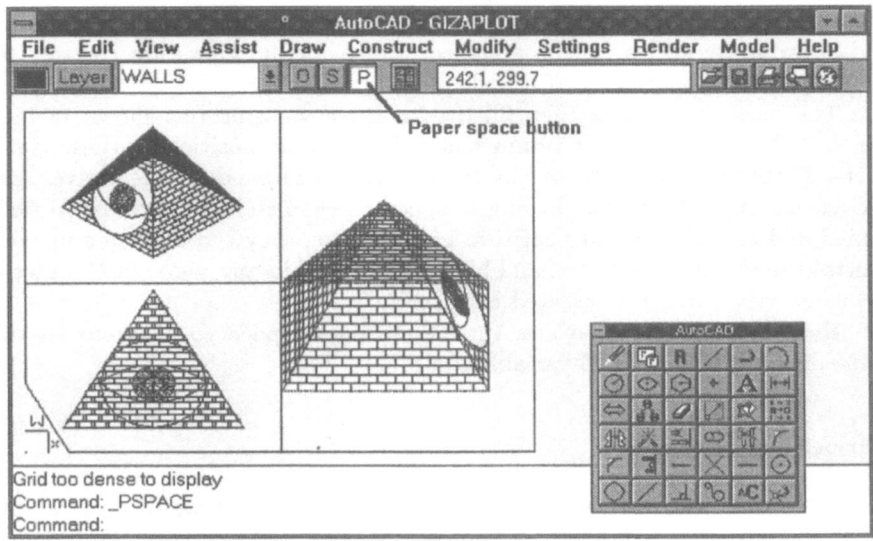

Figure 9.14 Starting Gizaplot

Layer name(s) for color 4 (cyan) <PSVP>: <**ENTER**>
?/Make/Set/...: <**ENTER**>

The LAYER command's "Make" option both creates a new layer and makes it the current one.

The manipulation of the metaviews frames involves four key stages. Firstly, all the viewports will be changed to the PSVP layer. Then the large viewport on the right will be erased to be replaced by two smaller ones. The first will be obtained by copying one of the existing ones, while MVIEW will be used for the second. Once the viewports have been created the appropriate view points and scales for the plan, front elevation and side elevation must be chosen. The final, stage is the fine tuning of the viewports to get the plan and elevation views aligned.

Now execute the changing of layers. Pick **Modify/Change** and then **Properties. . . .**

Command:_ddchprop
Initializing. . .**DDCHPROP** loaded
Select object: **W**
First corner: **0,0**
Other corner: **420,297**
3 found

Select objects: **<ENTER>**

The Change Properties dialogue box will then appear. Pick the **Layer...** button and then select the layer **PSVP**. The frames of the viewports should have changed to cyan which is the color of the PSVP layer.

Note that the selection procedure above only found entities that had been created in paper space. The pyramid was left alone. Thus if you wish to change anything relating to the pyramid you must go to model space. This applies to the VPOINT as well as the entities of the pyramid. If you want to modify the location, size or appearance of the viewport frames then you must go to paper space. If your window selection didn't find any objects then check whether you are in paper space or not. When in paper space the P button should be bright, or P should appear on the status line. If in doubt type the command **PSPACE** and try again.

Now erase the right hand viewport entity.

Command: **ERASE**
Select objects: **410,150**   (Point on right hand edge)
1 found
Select objects: **<ENTER>**

To fill the space left by this deletion, first copy the top left viewport from A to B as shown in Figure 9.15.

Command: **COPY**
Select objects: **10,200**   (Any point on the top frame will do)
1 found
Select objects: **<ENTER>**
<Basepoint or displacement>/Multiple: **10,285**   (A)
Second point of displacement: **170,180**   (B)

Everything about the viewport is copied including the VPOINT of the pyramid. This newest viewport overlaps the two others. In earlier versions of AutoCAD and when TILEMODE was 1 ie ON viewports could not overlap but could be arranged side by side like tiles – hence the term *tilemode*.

The final metaview shown in Figure 9.16 could also be created by copying, since it is the same size as the others. However, as a refresher on the MVIEW command we will create it from scratch. Pick **View/Mview** followed by **Create Viewport**.

Command: **MVIEW**
ON/OFF/Hideplot/Fit/2/3/4/Restore/<First point>: **210,147.5**
Other corner: **410,285**

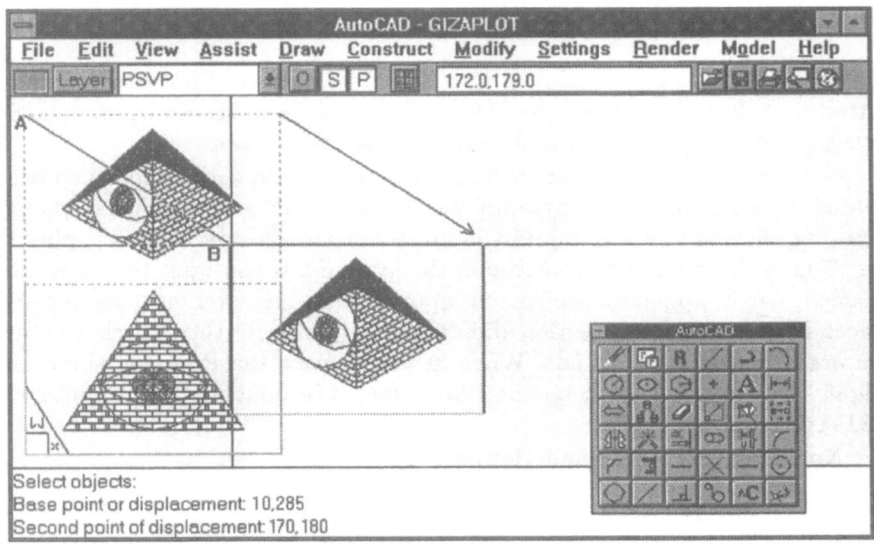

Figure 9.15 Copying a metaview

Initially, the new viewport will display the same image as the currently active viewport in MODEL SPACE. The next stage involves going to model space, making each viewport in turn active and choosing the viewpoint and scale of the view. At the end of this stage you screen should have four viewports, as in Figure 9.16.

Command: **MSPACE**

The upper right viewport should be outlined to indicate that it is the active one. When you move the cursor the cross hairs should only appear in this viewport. Elsewhere on the screen it appears as an arrow. If any other viewport is the active one, move the arrow cursor to the top left viewport and press the pick button.

Change the VPOINT to generate a side elevation. Pick **View/Set View** with **Viewpoint/Set Vpoint** or type **VPOINT** at the command prompt. The vpoint is the position of an observer looking towards the origin. The side elevation can be got by looking from the point (-1,0,0) towards (0,0,0). Then use the Zoom, Scale-XP option to set a 1:5 scale. The XP stands for "times paper space".

Command: **VPOINT**
Rotate/<View point> <current>: **-1,0,0**

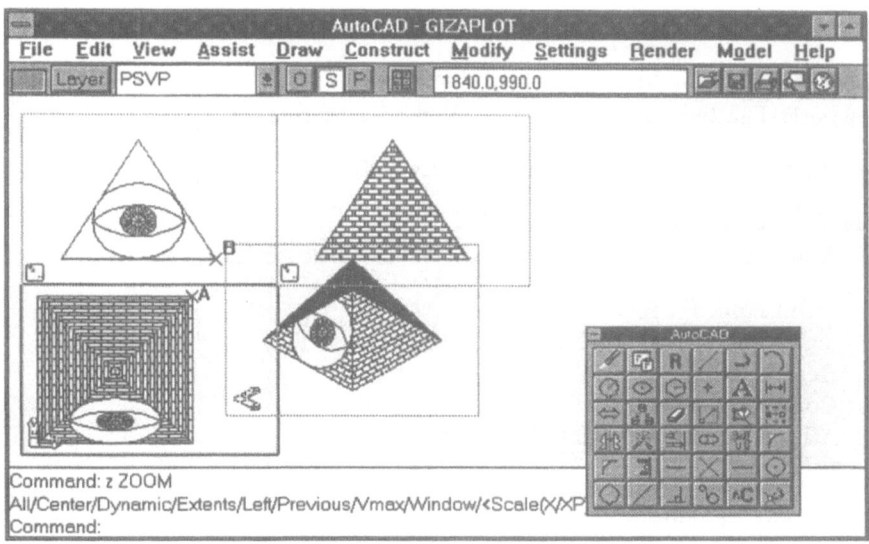

Figure 9.16 The right viewpoint

Regenerating drawing.
Command: **ZOOM**
All/.../Window/<Scale (X/XP)>:**.2xp**

This view should now look like that in Figure 9.16. To get the other views as shown do the following. Make the upper left viewport the active one. Move the cursor into the other viewport and click the cursor. It should change from an arrow shape to the usual cross hairs. This view will now be changed to show the front elevation of the pyramid.

Command: **VPOINT**
Rotate/<View point> <1.0,-1.0,1.0>: **0,-1,0**
Regenerating drawing.
Command: **ZOOM**
All/.../Window/<Scale (X/XP)>:**.2xp**

Having got things down to size, the next thing is to make the all seeing eye visible again. The reason it is not clear at the moment is because of all the bricks belonging to the back wall. To make the bricks disappear as shown in Figure 9.16 we can freeze the WALLS layer for this viewport. This has to be done via the Layer Control dialogue box. Pick **Settings/Layer Control....** In the dialogue box pick the line:

WALLS   On . . .. white   CONTINUOUS

Then click the **Cur VP: Frz** button. This freezes the layer only in the current viewport. The line in the dialogue box should change to:

WALLS   On . . C. white   CONTINUOUS

Pick **OK** to execute the change.

Now make lower left viewport active and execute the PLAN command.

Command: **PLAN**
<Current>/Ucs/World: **W**
Regenerating drawing.
Command: **ZOOM**
All/.../Window/<Scale (X/XP)>:**.2xp**

This leaves the display as shown in Figure 9.16. The job is not quite finished yet, though. The plan view is out of line with the front elevation in the top left view and the isometric could be better placed.

The first part of the fine tuning stage is done in paper space. You need to move the bottom left viewport a shade to the right. Then you will set up the hidden line removal for plotting the isometric view.

Switching is done with the **PSPACE** command. The shifting of the viewports is done with AutoCAD's standard **MOVE** command. Hidden line removal is one of **MVIEW**'s options.

Command: **PSPACE** or pick the P button on the tool bar.

In the DOS version, the "P" should reappear on the status line. The coordinate read-out should reflect the paper space limits.

To align the plan view with the front elevation you need to move the lower left viewport to the right. The magnitude of the displacement can be got by using object snap to pick up the points A and B in Figure 9.16. The coordinate filters, .x, and .yz, are also used.

Command: **MOVE**
Select objects: **100,10**   (Point on edge of bottom left viewport)
1 found
Select objects: **<ENTER>**
Base point or displacement: **intersec** of Pick point A   (near 145,140)
Second point of displacement: **.x** of **intersec** of Pick point B
(need YZ) **intersec** of Pick point A again.

This viewport should now appear shifted to the right as shown in Figure 9.17. A point to note about the previous manipulation is that you were able to snap

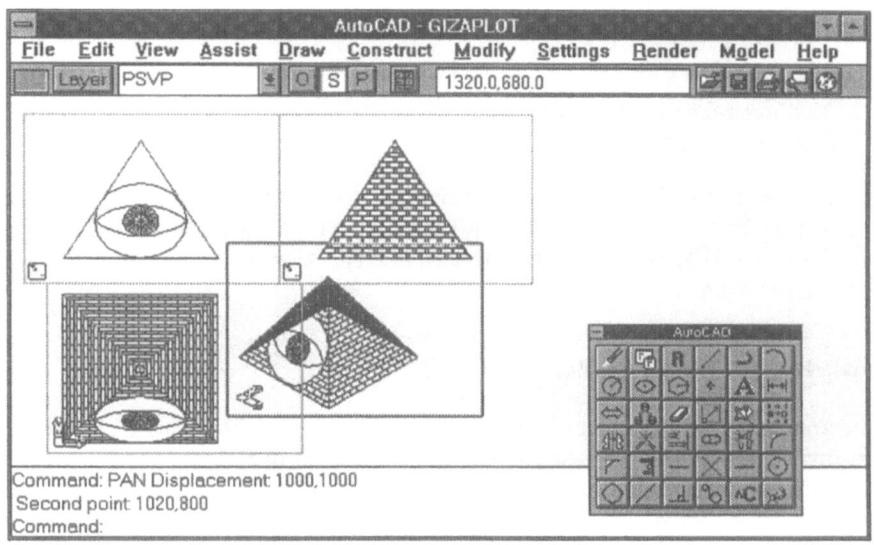

Figure 9.17 Fine tuning

to objects that have been defined in model space even though the paper space was active. However, it is not possible to make any physical alterations to the pyramid when paper space is active.

If you set "Hideplot" to "on" then a hidden line removal will be done in this view when the drawing is plotted. This is one of the options of the **MVIEW** command. You select a particular viewport by picking one of its edges.

Command: **MVIEW**
ON/OFF/Hideplot/Fit/2/3/4/Restore/<First point>: **H**
ON/OFF: **ON**
Select objects: **370,100**   (Point on edge of bottom right viewport)
1 found
Select objects: **<ENTER>**

The final adjustment is to the isometric view. This needs to be shifted a bit down and to the left. This can be done in either of two ways. Firstly, you could MOVE the viewport as before. Secondly, you could go back to model space and using the PAN command shift the pyramid within the viewport. For variety try the latter.

Command: **MSPACE**

Make the lower right viewport the active one and do the PAN.

Command: **PAN**
Displacement: **1000,1000**   (Corner of pyramid)
Second point: **1020,800**

This last point was found by trial and error to give a reasonable display. Picking points in isometric views is always a bit risky. The final stage in this exercise is to draw the title block and margins, make the viewport edges invisible and plot the drawing.

Cross referencing drawing files

Now that all the views have been set up we can add in the title block and margin. This will be done using a feature introduced in Release 11, namely, the ability to cross reference drawings. Prior to Release 11 the only way to import a second drawing into the current one was to use the INSERT command. This physically copied all the entities of the second drawing in the form of a block. Now, a command called **XREF** can be used to link the current drawing with a previously defined one.

The main advantage of using XREF over INSERT is that the entities of the referenced drawing are not copied into the current one. They are simply displayed at the same time. This helps to keep down the size of the current drawing file. A second advantage is that any modifications made to the referenced drawing will automatically be made to the referring one.

While still in paper space we will XREF the drawing TITLE-A3. As XREF will be executed while in paper space, the margins and title box also become part of the paper space. It will also be attached to the current layer. Before generating the final plot, we must make the edges of the viewports invisible. To do this you must make the PSVP layer invisible. Go to paper space and change to layer 0 and freeze PSVP.

Command: **PSPACE**
Command: **LAYER**
?/Make/Set/...: **S**
New current layer <PSVP>: **0**
?/Make/Set/.../Freeze/Thaw: **F**
Layer name(s) to freeze: **PSVP**
?/Make/Set/...: **<ENTER>**

You can now perform the XREF. Pick **File/Xref/Attach...** from the menu bar.

Command: _xref

?/Bind/Detach/Path/Reload/<Attach>: _attach

The Select File to Attach dialogue box will then appear. Pick **title-a3.dwg** from the list of files followed by **OK**.

Xref to Attach: ~~
Attach Xref TITLE-A3: TITLE-A3.DWG

The remainder of the dialogue is similar to that for block insertion.

Insertion point: **0,0**
X scale factor <1>/Corner/XYZ: **<ENTER>**
Y scale factor (default=X): **<ENTER>**
Rotation angle <0.00>: **<ENTER>**

Of the other options in the XREF command line "detach" allows you to remove a cross reference that was previously attached. The "bind" option will convert a XREF drawing into a conventional AutoCAD block. "Path" is used if the referenced drawing is been moved to another directory or if its filename is changed. When you enter the AutoCAD drawing editor it automatically reloads any XREF's. The "Reload" option does the same within an editing session. This might be useful if a group of people are working in a series of interconnected drawings. Finally, the "?" will give you a list of all the external references in the current drawing.

Dependent blocks

If you look at the list of blocks you will find that TITLE-A3 is defined as an external reference block. The another block TITLE-A3|TTEXT is also defined as a dependent block. This is, of course the TTEXT block that was defined as part of the TITLE-A3 drawing earlier. Even though the block appears on the list it is not accessible in its current form.

Command: **BLOCK**
Block name (or ?): **?**
Defined blocks.

TITLE-A3	Xref: resolved
TITLE-A3|TTEXT	Xdep: TITLE-A3

User blocks	External References	Dependent Blocks	Unnamed Blocks
0	1	1	4

In order to be able to use the TTEXT block it has to be bound to the current drawing. This can be done by binding the whole of TITLE-A3 which converts it to a conventional block and also converts the nested block. It can also be done by just binding the TTEXT block and leaving the rest of TITLE-A3 as an XREF. A special command, **XBIND** does the latter. One of the difficulties in doing this is that some keyboards do not have the "|" character. You should be able to generate it from its ASCII code 124. To do this hold down the **ALT** key and type **124** on the numeric keypad at the right of the keyboard.

Command: **XBIND**
Block/Dimstyle/Layer/LType/Style: **B**
Dependent Block name(s): **TITLE-A3|TTEXT**
Scanning...
1 Block(s) bound.

At this point the block will become a conventional part of the current drawing. Its name will change to "TITLE-A3$0$TTEXT". The | is changed to 0. This new block can now be inserted. You may have to input the title and name using the dialogue box.

Command: **INSERT**
Block name (or ?): **TITLE-A3$0$TTEXT**
Insertion point: **0,0**
X scale factor <1>/Corner/XYZ: **<ENTER>**
Y scale factor (default=/x): **<ENTER>**
Rotation angle <0.00>: **<ENTER>**
Enter the drawing title: <No title>: **Pyramid of Giza**
Enter your name: <Anonymous>: **M.Y. Name**

Your screen should now look like the plot shown in Figure 9.18 without the hidden line removal. You can plot it now or use the Windows File Manager described in the next section.

Before using the Windows printing facilities you must save the drawing. It is not recommended to try and print a file from Windows while the file is open in AutoCAD. Pick **File/Exit AutoCAD** and then pick **Save Changes**.

A note on XREF files

The XREF'fed file cannot be altered from inside the referring file. Thus you cannot change the contents of the drawing TITLE-A3 from inside the drawing "GIZAPLOT". If you write your name within the title box, then that text will be part of the current drawing only. However, you can use object snap to points in the XREF'fed drawing.

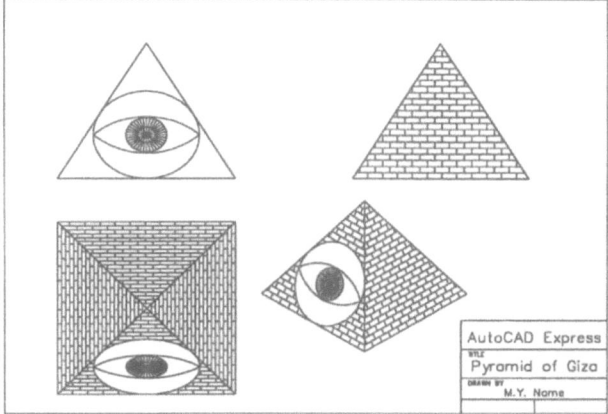

Figure 9.18 The final plot of Giza

The referenced drawing will be associated with whatever is the current layer when the XREF command is issued. Above, the title box and margins are attached to layer 0. However, it is not as simple as all that. A new layer, called "TITLE-A3|MARGIN" will also be created automatically. The visibility of the margins etc is controlled by both layer 0 and TITLE- A3|MARGIN. If either is frozen then the margins will be invisible.

When you exit a drawing that contains XREFs the information from the referenced file is deleted from the current one. Only the actual reference is retained. This is used the next time you edit the drawing to resolve the XREF and display the drawing in full.

You can only xref a drawing that has been created in model space. Paper space entities are not copied.

Printing from Windows

Clicking the File Manager icon from the "Main" program group in Windows gives a list of files similar to that shown in Figure 9.19. It is relatively easy to navigate around the disk's directory structure. All you do is double click the name of the directory in the left hand part of the window to reveal the files and sub-directories.

To plot an AutoCAD file you just need to locate the appropriate ".dwg" file. In this case click *once* on the file **gizaplot.dwg**. Then pick the **File** pull down menu followed by **Print**. A pop-up box will then ask you to confirm that you want to print the file. Pick **OK**.

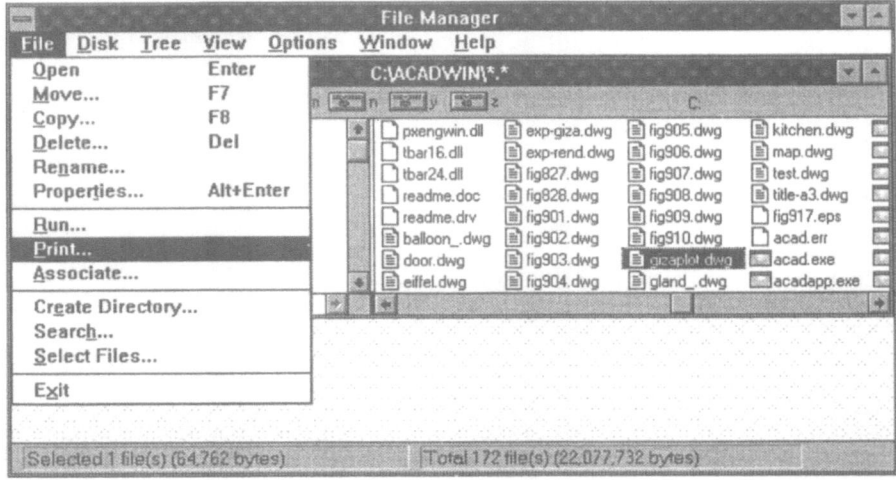

Figure 9.19 File Manager

This has the effect of starting up AutoCAD, loading that drawing file and executing the **PLOT** command. You will get the Plot Configuration dialogue box as before. Set all the parameters as required. Note that GIZAPLOT was in paper space when saved and so can be plotted at a scale of 1 plotted mm equals 1 drawing unit if the plotter has A3 capabilities. As AutoCAD plots this file it processes each viewport, in turn, removing hidden lines in the one where "hideplot" was set "on".

When the plot has finished you will be left in the AutoCAD editor. Pick **File/Exit AutoCAD** to get back to the file manager. Note that if you double click on a ".dwg" file in the File Manager it will start up AutoCAD and load that file.

Windows system printer

Both the DOS and Windows versions of AutoCAD come with a host of printer and plotter drivers. These are translator programs which allow AutoCAD to output the plot in the correct format for the device. With AutoCAD for Windows, one of these is the Windows System Printer. This is the printer or plotter that Windows itself is configured to work with. Autodesk recommend that you use the appropriate device driver rather than use the Windows System Printer wherever possible. In particular Autodesk's ADI drivers are quicker and more direct than printing via Windows.

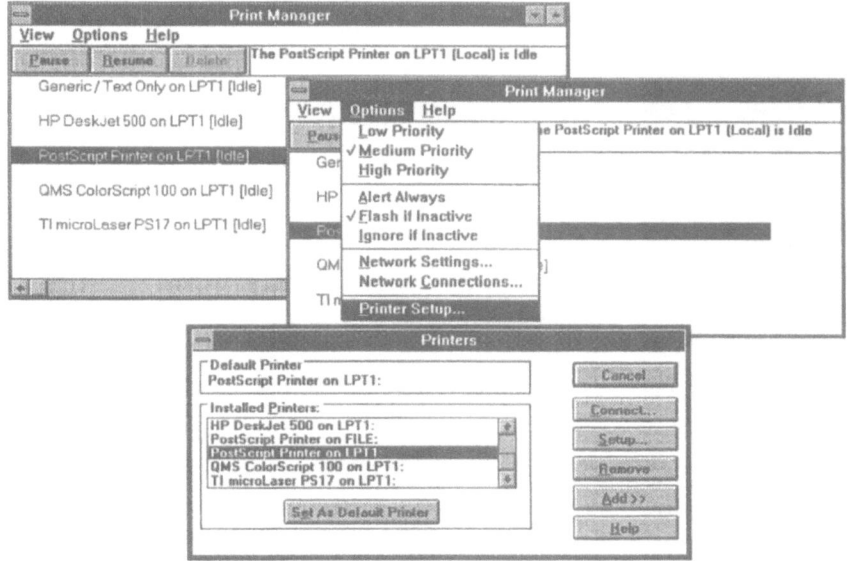

Figure 9.20 Windows Print Manager

However, printing via Windows does have some benefits. One is that you may already be using Windows for printing from other packages. The main advantage is that you can use the Windows Print Manager program to direct the output to different printers. This can be especially useful when printing over a network.

Figure 9.20 shows the main screens of the Print Manager. When you pick the Print Manager icon from the "Main" program group you get a list of the printers currently available. It will also show any files you may have in the print queue. Picking **Options** allows you to set priorities, connect to networked printers and to modify the printer setup. This allows you to set the default Windows System Printer setting.

As long as AutoCAD has been configured for the Windows System Printer you can access it when plotting. Figure 9.21 shows the route. Picking **Device and Default Selection** from Plot Configuration is followed by picking **Windows System Printer**. Then, if you pick **Change Device Requirements...** you get the Print Setup box. This is the standard Windows setup that you get with any application.

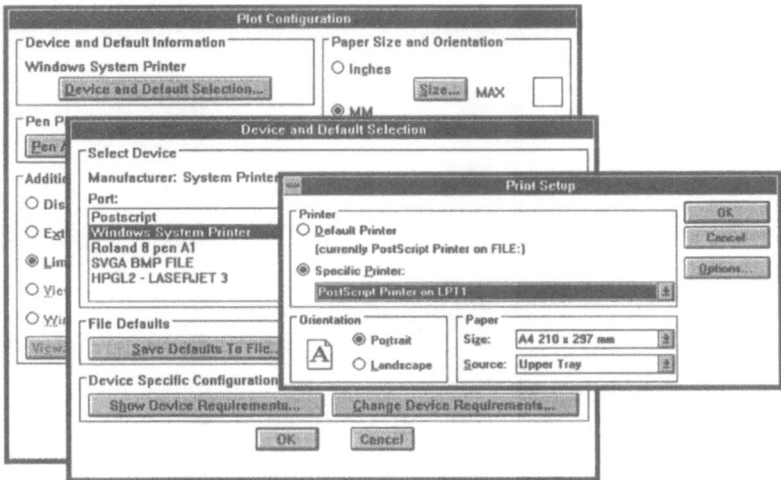

Figure 9.21 Windows System Printer in AutoCAD

The softcopy

Safety first

In any drawing office, security and archiving are given great priority. This
must be extended when drawings are stored on magnetic media. Some users
will have access to sophisticated CAD management programs that take care
of this. Many will just be relying on sensible practices.

The first sensible practice is to save the drawing regularly. In particular
the drawing should be saved before "hazardous" operations such as plotting
or using HATCH. AutoCAD provides one backup copy in the same directory
as the drawing. You, the sensible user, should have another backup on another
disk. A further safe copy should be stored in a separate and safe location. If
one copy gets corrupted make sure it is replaced immediately, don't put it off.

Large volume archiving is best achieved by copying to a tape storage
device. This normally requires specific hardware attached to your computer.
Another device suitable for mass storage is the WORM (Write Once Read
Many) laser disk system.

If you are archiving drawings that contain xref's then it is recommended
that you bind the referenced drawings to the archived one using XBIND. Alternatively, you can archive all the xref's as separate files along with the drawing
that uses them. You are also recommended to bind all references when creating
DXF or IGES files for interchange with other CAD systems.

Copying to other Windows applications

A great advantage of AutoCAD for Windows is that it opens up all the possibilities of that environment. Not only can you resize the drawing window and
simultaneously run other applications but you can use Windows to communicate between AutoCAD and these applications. This means that you can copy
text and drawing notes from your Windows wordprocessor to AutoCAD and
copy drawings into your wordprocessor.

To copy text into an AutoCAD drawing you must use the Windows clipboard. When there is something in the clipboard, AutoCAD's **Edit** pull-down
menu allows you to **Paste Command**. This should only be picked when the
DTEXT command is running and prompting for text. You could use the Windows WRITE program to generate the text. Select a portion of the text that
doesn't contain any blank lines and copy it to the clipboard. CTRL and C
usually does this or **Edit/Copy** from the pull-down menu. Then switch to
AutoCAD, either using the ALT Tab toggle or CRTL and Esc. In AutoCAD
begin the DTEXT command.

Command: **DTEXT**
Justify/Style/<Start point>: pick where you want the text inserted.
Height<current>: input the desired height
Rotation angle <0.00> **<ENTER>**
Text: pick Edit/Paste Command

Now when you pick **Edit/Paste Command** the text will appear here and on
the drawing.

This text then becomes a normal AutoCAD text entity. Note that there
is no active link between the original wordprocessor file and the drawing. The
clipboard merely did a copy operation. If you change the text in the original
file it will have no new effect on the AutoCAD text.

It is possible to copy an AutoCAD image to any other windows application. This can be done using **Edit/Copy Image** or **Edit/Copy Vectors**.
The COPYIMAGE command creates a bitmap image of a selected part of the
graphics screen. This is copied to the clipboard and can be pasted into any
other Windows application. Bitmap images can only be edited using programs
such as PAINTBRUSH.

Picking "Copy Vectors" is a bit more sophisticated. This allows you to pick
individual entities for copying to the clipboard. Furthermore, these are copied
in Windows WMF vector format and can be pasted into another application
including AutoCAD. This gives a rapid method of copying part of a drawing
to another. Many Windows applications can read WMF files. Those that can't
will usually read a bitmap image.

Object linking and embedding

Many Windows applications support Object Linking and Embedding (OLE). This allows you to create active links between applications. AutoCAD supports OLE as a server application. To use OLE you need the other application to support it as a client. This means that AutoCAD is the source and the other application receives the drawing. For example, you can link or embed an AutoCAD drawing in a WRITE or Microsoft Word document but not vice versa. The difference between OLE and the simple cut and paste done above is that when a drawing is embedded in a document it can be edited using AutoCAD or the server application. The difference between linking and embedding is subtle. When an AutoCAD link is established, any changes to the original drawing will be applied in the document by picking UPDATE from the Edit/Link menu. Only one copy of the drawing file exists. When you embed a drawing the original AutoCAD drawing is *copied* into the document.

The following example links the drawing BALLOON.DWG to a document created with WRITE. Start up WRITE by picking the icon from the Accessories program group. Type in some text and position the cursor where you wish to insert the diagram. Then pick **Edit/Insert Object**... from the menu bar. Pick **AutoCAD Drawing** from the Insert Object pop-up. This will start up the AutoCAD editor. Pick **File/Open** and open the drawing "balloon.dwg". Do whatever edits or zooms you wish. When you are happy with the display save it using the VIEW command. Then use VIEW, Restore to make this active or restore one of the other saved views. It is this currently active view that will be displayed in the document. Then pick **Edit/Copy Link** from the menu bar followed by **File/Exit AutoCAD**. You must save the changes or the link will not work.

Exiting AutoCAD lands you back in the WRITE document. Now pick **Edit/Paste Link** and presto. Note that the view does not work in the same way as for plotting, the whole display is shown. If you did not save and restore a view AutoCAD will automatically do one for the current display. Thus even if *you* did not change anything in the drawing it is essential to save the changes to the view information in AutoCAD.

You can edit the image of the tower by double clicking on it in the document. This opens AutoCAD once more with the drawing file loaded. Do some edits or change the view definition. Then exit AutoCAD saving the changes. The modification will automatically be updated. Remember, it is the original view name that was used with AutoCAD's Copy Link that controls what will be displayed. You can cancel the link in WRITE by picking **Edit/Links** and Cancel from the Links pop-up. This pop-up also allows you to edit the drawing or change the link to a different drawing.

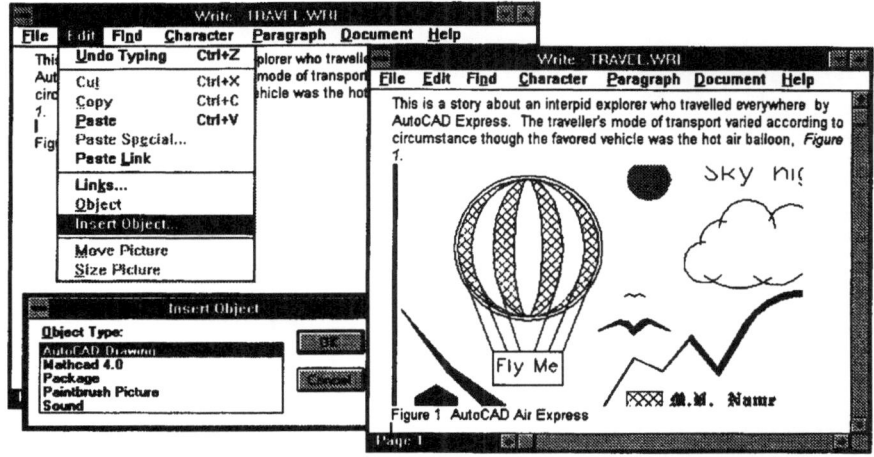

Figure 9.22 Object linking

Summary

This chapter has covered the procedures for producing hardcopy output of AutoCAD drawings. It has also introduced a number of Windows functions for copying AutoCAD drawings to other applications.

You should now be able to

Set up the printer/plotter parameters.
Plot multi-colored drawings on pen plotters.
Direct output to a file.
Copy plot files to the COM1 serial port.
Use **MVIEW** to make overlapping viewports.
Make layers frozen in individual viewports.
Plot multiple viewports with hidden line removal.
Use cross referenced files.
Bind dependent blocks.
Print AutoCAD drawings from the File Manager.
Set up the Windows System Printer.
Copy to the Windows clipboard.
Link and Embed an AutoCAD drawing in another application.

Appendix A CONFIGURATION

General

The configuration of AutoCAD to work on your particular hardware involves a few stages. Firstly, you must set the computer's working environment and then set up the AutoCAD environment in Windows. Finally, you must tell AutoCAD about the various pieces of hardware it has to communicate with. The chief difficulty with describing system configurations is that they depend on many factors. Thus, comprehensive coverage of all the possible AutoCAD configurations is not feasible in a short book. The major considerations are outlined below, though slight modifications may be required for your own installation. If in doubt, you should consult the *AutoCAD Interface, Installation and Performance Guide – Windows*.

The DOS environment

The computer's working environment on PCs is controlled through two files, CONFIG.SYS and AUTOEXEC.BAT. The first file contains parameters telling your computer how to utilize its memory and may also contain some default settings for its operation. AUTOEXEC.BAT is a list of DOS commands which are automatically executed when you turn the computer on. Both of these files are ASCII text files and can be created with EDIT or a wordprocessor. The sample files given below are the ones I use on my own computer. They are set up for running Windows. No extra settings are required for AutoCAD for Windows.

HAZARD WARNING! It can be dangerous to tinker with the autoexec.bat and config.sys files. The sample files given here may conflict with other software on your system. If in doubt, consult your Micorsoft DOS and Windows manuals.

DEVICE=C:\DOS\HIMEM.SYS	PATH C:\WINDOWS;C:\DOS;C:\;
DEVICE=C:\DOS\EMM386.EXE NOEMS	PROMPT PG
DOS=HIGH	SET TEMP=C:\TEMP
FCBS=4,0	REM Run Windows
STACKS=9,256	WIN
COUNTRY=044,,C:\DOS\COUNTRY.SYS	
FILES=40	
DEVICE=C:\DOS\SETVER.EXE	
Example CONFIG.SYS FOR DOS 6.0	Example AUTOEXEC.BAT

In earlier versions of AutoCAD the operating environment was set up in the autoexec.bat file. However, with AutoCAD for Windows this is now done in a file called "acad.ini". This file tells AutoCAD where to find the support files, where to write the memory pager etc... You can access this file and alter the environment from the AutoCAD drawing editor.

Start AutoCAD and pick **File/ Preferences** from the menu bar. This was briefly described in Chapter 2, Figure 2.4. The Preferences dialogue box allows you to configure the AutoCAD Graphics Window, the text window and modify the environment settings. The buttons on the bottom of the dialogue box allow you to customize the colors and fonts used by Windows. Pick the **Color...** button or **Fonts...** to do this. Note these are colors and fonts for the window and not for the AutoCAD drawing. Picking the **Environment...** button gives another dialogue box showing the Support directories, the Help file location, the Page file directory and the Log file name and location.

The support directories will be searched by AutoCAD whenever a font file or block file is needed. If you created a "SYMBOLS" sub-directory in Chapter 6 then it should be added to the list of support directories. Each path in the list is separated by a semi-colon. Make your changes and pick **OK**. To save any changes permanently pick **Save to ACAD.INI** in the Preferences dialogue box, followed by **OK**.

Configuring AutoCAD

The second stage of setting up AutoCAD on your computer is to tell the program about the hardware it has available to it. You must specify what kind of video display unit is installed, what type of printer and plotter and the make of your digitising or pointing device. This is done in AutoCAD by picking **File/ Configure** from the menu bar. The text window will then open displaying the current configuration, if one exists. For example, the configuration might appear as:

Current AutoCAD configuration
    Video display:
      Windows Accelerated Display Driver ADI 4.2 – by Autodesk, Inc
        Version: 12
    Digitizer:
      Current System Pointing Device
    Plotter1:
      Postscript device ADI 4.2 – by Autodesk, Inc
        Model: 300 dpi
        Connected to device: LPT1
        Version: A.1.80
    Plotter2
    Press RETURN to continue: **<ENTER>**

Pressing the ENTER key then shows the other plotter configurations. When all of the current configuration screens have been displayed AutoCAD moves to the Configuration menu:

Configuration menu
    0.   Exit to main menu
    1.   Show current configuration
    2.   Allow detailed configuration

    3.   Configure video display
    4.   Configure digitizer
    5.   Configure plotter
    6.   Configure system console
    7.   Configure operating parameters

    Enter selection: <0>:

To select an option, type the appropriate number followed by <ENTER>. The options 0–2 don't cause any actual changes to the configuration and so are separated from the other selections. The rest are used to alter the set-up of individual devices.

    Enter selection: <0>: **2**
    Do you want to do detailed device configuration? <N> **Y**
    Additional questions will be asked during device configuration.
    Press RETURN to continue: **<ENTER>**

You must choose this option if you wish to alter the DOS "MODE" parameters for peripherals connected to the input/output ports on your computer. You will require some technical knowledge to do this though AutoCAD does offer sensible defaults. You should also use this option to tell AutoCAD to which port (Serial = COM1 or COM2 or Parallel = LPT1, LPT2 or LPT3) your plotter, printer and digitizer are connected. Consult the operating manual for your mouse, plotter or printer when using this option.

When you have made all the desired modifications you can exit the configuration menu by selecting "0". You will be asked if you want to keep the changes or to discard them.

Enter selection <0>: <ENTER>
If you answer N to the following question, all configuration

changes you have just made will be discarded.
Keep configuration changes? <Y> <ENTER>

Altering devices

When you select any of the options 3 to 6 you will be told what your current device and its settings are. You then have the choice of simply altering the settings or of choosing a new device. Depending on the type of device to be configured, AutoCAD displays a list of makes and models that are acceptable. You simply type the number given for your device and answer a few questions about it. If your device does not appear on the list you should consult the device's operating manual. Many smaller manufacturers build their equipment to emulate better known makes. For example,is emulated by many makes of video display unit, while Hewlett-Packard is the *de facto* standard for plotters.

Configuring the video display unit.   As an example, to change the display configuration, you might do the following:

Enter selection: <0>: **3**
Your current video display is: Windows Accelerated Display Driver ADI
Do you want to select a different one? <N> **Y**

You will then be presented with a list of the devices supported by AutoCAD. In the standard set up there are only two similar devices. More may be displayed if you have third party ADI drivers with your graphics card.

Available video displays:
  1. Windows Accelerated Display Driver ADI 4.2 – by Autodesk, Inc
  2. Windows driver – by Autodesk, Inc
Select device number or ? to repeat list <1>: **1**

The accelerated driver uses more system memory but is much faster then the other one. If you select the Windows Accelerated Display Driver, a pop-up menu will appear. Activate both the **Display-list** and **GDI bypass** by picking the boxes. They are active if an X appears in the box. The display-list option gives faster graphics but uses a bit more memory. The GDI bypass allows AutoCAD to bypass some of the windows software. Again this speeds things up. Pick **OK** to continue with the configuration. You will be given the opportunity to correct any aspect ratio distortion of the screen image.

If you have previously measured the height and width of
a "square" on your graphics screen, you may use these
measurements to correct the aspect ratio.
Would you like to do so? <N> **Y**

The aspect ratio problem is caused by the screen pixels being taller than
they are wide resulting in circles that look like ellipses and squares that look
like rectangles. If this happens on your screen, measure the width and height
of the object that should be square with a ruler. You can then use these
values as inputs for AutoCAD's configuration. The units (mm or inches etc)
of measurement don't matter as long as they are the same for width and height.

Width of square <1.0000>: Give measured value
Height of square <1.0000>: Give measured value

After giving the square sizes or if you answered "N" to the previous question,
you will then be returned to the Configuration menu.

Configuring the digitizer The term "digitizer" covers most forms of pointing
device from the relatively cheap mouse and joystick to tracker balls, space
balls and A0 digitising tablets. The configuration procedure is quite simple.
You enter selection number 4 and choose your device's number from the list
displayed. You may also tell AutoCAD where the device is conected if you
have previously requested the "Allow detailed configuration".

To select the Windows pointing device, you might do the following:

Enter selection: <0>: **4**
Your current digitizer is: Summagraphics MM Series v2.0 ADI 4.2 −
Do you want to select a different one? <N> **Y**
Available digitizers:
   1. Current System Pointing Device
   2. CalComp 2500 and 9100 Series ADI
   etc.
Select device number or ? to repeat list <>: **1**

The Configuration menu then reappears. If you had selected a different device
and having previously selected "Allow detailed configuration" you would get
the extra prompts:

Connects to Asynchronous Communications Adapter port.
Standard ports are:
COM1
COM2
Enter port name, or address in hexadecimal <COM1>: **COM2**

Different makes of digitizer result in different prompts. In general the questions
are self-explanatory and the defaults usually work quite well.

Configuring the plotter.   The main difference between plotters and the other devices is that AutoCAD allows you to store multiple plotter configurations. Then when it comes to plotting a drawing, as in Chapter 9, you can choose which plotter to use. However, you can only choose from those already configured. Both plotters and printers now come under the plotter configuration menu. Entering **5** at the Configuration Menu gives the Plotter Configuration Menu. The sequence below is to add a HPGL-2 laser printer.

Enter selection: <0>: **5**
Your current plotter is: System Printer ADI 4.2 – by Autodesk, Inc
Description: Windows System Printer
Plotter Configuration Menu
    0.   Exit to configuration menu
    1.   Add a plotter configuration
    2.   Delete a plotter configuration
    3.   Change a plotter configuratoin
    4.   Rename a plotter configuration
Enter selection, 0 to 4 <0>: **1**

The procedure to add a plotter is much the same as or selecting the digitizer. You are shown a list of available plotters.

Available plotters:
  1. None
  2. AutoCAD file output formats (pre 4.1) by – Autodesk, Inc
  ⋮
  9. Hewlett-Packard (HP-GL/2) ADI 4.2 by – Autodesk, Inc
  etc.
Select device number or ? to repeat list <7>: **9**

The following prompts are specific to this device. Other devices will have their own set of questions to be answered.

Supported models:
  1. LaserJet III
  etc.
Enter selection, 1 to 6 <1>: **1**
Select the number of copies of each plot, 0 to 99 <1>: **1**
 Paper trays:
  ----
  1. Manual feed
  2. Letter
  3. A4 Sheet
  4. Legal
Select paper tray currently installed, 1 to 4 <2>: **3**
Is your plotter connected to a <S>erial, or <P>arallel port? <P> **P**
Enter parallel port name for plotter or . for none <LPT1>: **LPT1**

Next, you are asked about how the plots will be generated. These parameters can all be altered at plot time. Finally, you must give a name for this configuration.

> Plot will NOT be written to a selected file
>
> Sizes are in Inches and the style is landscape
>
> Plot origin is at (0.00, 0.00)
>
> Plotting area is 11.20 wide by 7.80 high (A4 size)
>
> Plot is NOT rotated
>
> Hidden lines will NOT be removed
>
> Plot will be scaled to fit the available area
>
> Do you want to change anything? (No/Yes/File) <N>: **N**
>
> Enter a description for this plotter: **HP Laserjet 3**

You are then returned to the Plotter Configuration Menu. Enter **0** to go to the main Configuration Menu.

> Enter selection, 0 to 4 <0>: **0**

Configuring the operating parameters.   Selecting option **7** from the configuration menu gives a further menu for various operating parameters. This allows you to specify the default prototype drawing name, plot filenames. If your computer is in a network you can specify the node name and plot spooler directory. Extra software is required to control the plot spool. This is a queuing system for plotting files.

Selecting sub-option 5 allows you to direct AutoCAD's temporary files to a specific disk drive or directory. When editing large drawings, AutoCAD uses disk space for storing parts of the drawing and information for the UNDO command etc. You can speed up network performance by specifying a local disk for the temporary files.

Option, 7, in the sub-menu allows you to set the time for automatic saving of the drawing, eg every 30 minutes. Item 10 in the list should only be used if there are problems with the HIDE command. The Release 12 algorithm for hidden line removal is much faster that previous ones. However, it sometimes yields strange results. The Release 11 algorithm is slower but more accurate. The final two items are for network use. File locking stops other network users opening your drawing files. The login name should be your own name. It will be used to tell other users who has locked the file.

> Enter selection <0>: **7**
>
> Operating parameter menu

  0.   Exit to configuration menu
  1.   Alarm on error
  2.   Initial drawing setup
  3.   Default plot file name
  4.   Plot spooler directory
  5.   Placement of temporary files
  6.   Network node name
  7.   Automatic-save feature
  8.   Full-time CRC validation
  9.   Automatic Audit after IGESIN, DXFIN, or DXBIN
 10.   Select Release 11 hidden line removal algorithm
 11.   Login name
 12.   File locking
Enter selection <0> **0**

Summary

Configuring AutoCAD for Windows is done in three separate ways. Firstly, there is the system configuration, secondly there is the AutoCAD environment and thirdly the device configuration. This Appendix is not meant to replace the *AutoCAD Interface, Installation andPerformance Guide – Windows* but can be used as a first port of call. For more difficult problems the reader is referred to the above manual.

Appendix B LINETYPES AND HATCH PATTERNS

General

The open nature of AutoCAD allows the user to define customised shapes, linetypes and hatch patterns. That is, should AutoCAD's already extensive library not be sufficient for you. Special text files must be created with a text editor. Once created, the linetypes and hatch patterns can be used like any of the standard ones.

Linetype definition

You may need to create your own linetype from time to time. For example, the LTSCALE may not give enough flexibility to show centre lines with varying dash lengths.

To create a new linetype you can modify an existing one in the ASCII text file called, "ACAD.LIN", or make a completely new file, MYLINE.LIN. The file type must be ".LIN". An example of a linetype with a dash length of 100 units, gap of 20, a dot, another gap of 20 all repeating is given below.

    *BIG_CENTERDOT,_____   ·   _____   ·   _____
    A, 25.0, −5.0, 0, −5.0

The first line begins with an asterisk, followed by the linetype name and a brief description which will appear as a drawing aid. The second line contains the actual definition. The "A" is an alignment code for AutoCAD and must appear as the first character in the line. Positive numbers indicate the length of dashes, negative numbers the gaps and a zero makes a dot.

To draw with this new definition you will have to load using the LINE-TYPE command, giving the name of the definition file.

Hatch definition

Hatch patterns are defined in a similar manner to linetypes. The standard patterns are stored in a file called ACAD.PAT. You can create your own hatch in a file called say, MYHATCH.PAT and then add them to the end of ACAD.PAT. The syntax for the pattern definition is given in the example below. Note, AutoCAD will only search the file ACAD.PAT for hatch patterns.

```
*V-pattern, A seriew of V shapes making a zigzag
315, 0, 0, 0.5, 0.5, 0.5, −0.5
45, 0.35, −0.35, 0.5, 0.5, 0.5, −0.5
```

The first line contains the name and a brief description. The second line starts with the angle of the line to be drawn, 315 degrees. This line starts from (0,0) and has a (delta x, delta y) repetition of (0.5,0.5). The last pair of numbers indicate that it is dashed with equal dashes and spaces. The third line is similar but its origin is at (0.35,−0.35).

To add this to the ACAD.PAT file use the DOS copy command:

COPY ACAD.PAT+MYHATCH.PAT ACAD.PAT

You can then use the pattern with the HATCH command.

Command: **HATCH**
Pattern (? or name/U,style): **V-PATTERN**
etc...

Summary

Simple linetype and hatch patterns are relatively easy to define. If you want to extend your experience of linetype and hatch pattern definition you are recommended to examine the two files, ACAD.LIN and ACAD.PAT, which come with AutoCAD.

SUBJECT INDEX